75
FSG

ALSO BY AMY LEACH

Things That Are

THE EVERYBODY ENSEMBLE

FARRAR, STRAUS AND GIROUX
NEW YORK

THE EVERYBODY ENSEMBLE

DONKEYS, ESSAYS, AND OTHER PANDEMONIUMS

AMY LEACH

Farrar, Straus and Giroux
120 Broadway, New York 10271

Printed in the United States of America
First edition, 2021

Illustrations by Na Kim.

Library of Congress Cataloging-in-Publication Data
Names: Leach, Amy, 1975– author.
Title: The everybody ensemble : donkeys, essays, and other pandemoniums / Amy Leach.
Description: First edition. | New York : Farrar, Straus and Giroux, 2021. | Includes bibliographical references.
Identifiers: LCCN 2021021362 | ISBN 9780374109660 (hardcover)
Subjects: LCGFT: Essays.
Classification: LCC PS3612.E21275 E84 2021 | DDC 814/.6—dc23
LC record available at https://lccn.loc.gov/2021021362

Designed by Gretchen Achilles

1 3 5 7 9 10 8 6 4 2

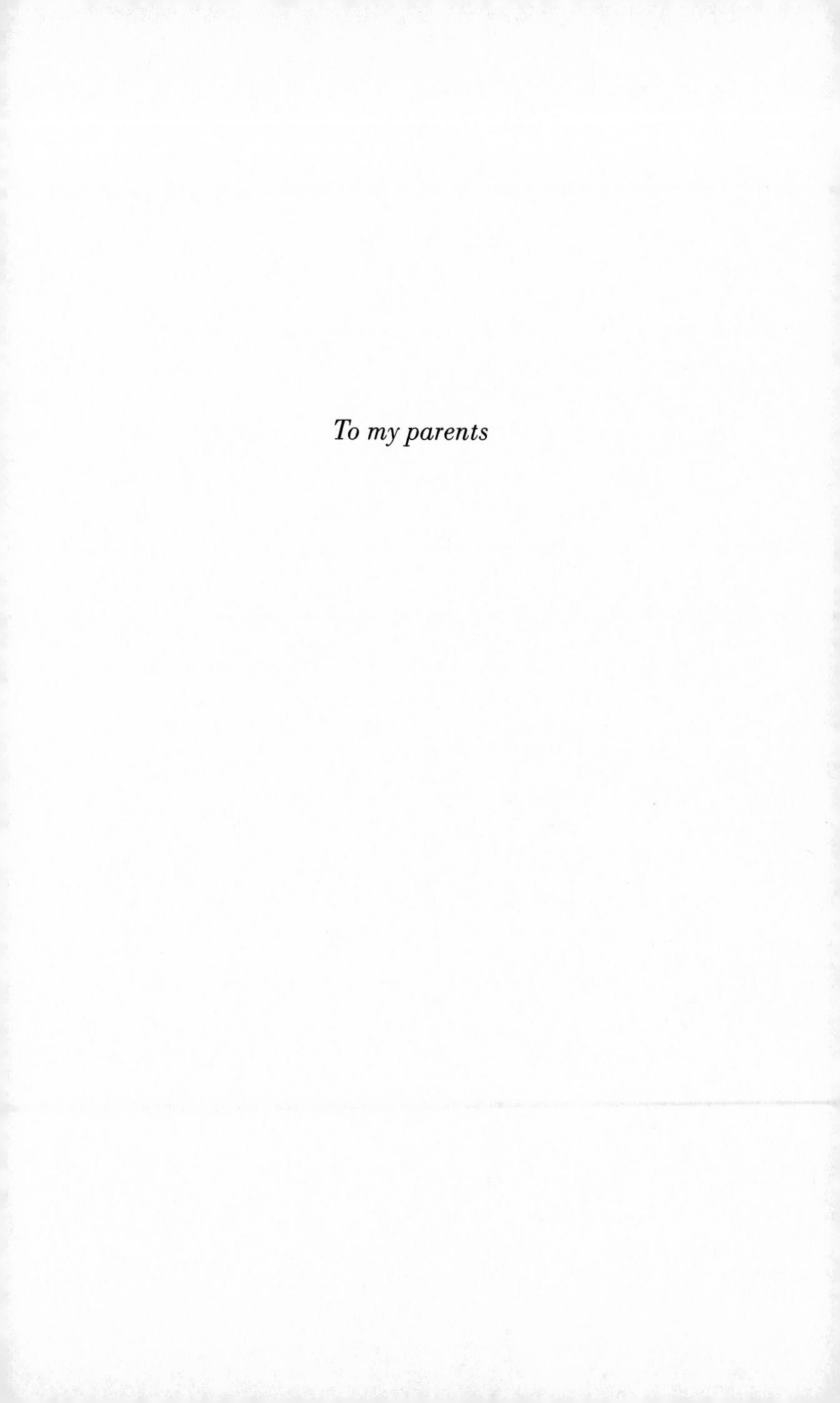

To my parents

But ask the animals, and they will teach you,
or the birds in the sky, and they will tell you;
or speak to the earth, and it will teach you,
or let the fish in the sea inform you.

—JOB 12:7–8

Flout 'em and scout 'em; and scout 'em and flout 'em:
Thought is free.

—WILLIAM SHAKESPEARE

CONTENTS

THE EVERYBODY ENSEMBLE

THE EVERYBODY ENSEMBLE

Welcome to the Everybody Ensemble! We're so glad you could make it for our concert tonight! We chose this location, where the Zambezi River empties into the Indian Ocean, so that aquatic and semiaquatic and land animals could all participate. The flapshell turtles didn't have far to travel, but we know that many of the rest of you have been traveling for months, even years, from Puducherry and the Grampian Mountains, from your bogs and boonies and cubicles, and we are grateful for all the trouble you took to get here. The trip would seem easiest for the birds, but of course they couldn't leave their eggs behind—and we see that some owls are still arriving, rolling their eggs around the mud puddles, stopping every several yards to sit on them and warm them up.

While they are settling in, let's talk about how you would like to be arranged. In a conventional choir the magic number is "four"—four sections corresponding to the four registers of the human voice—soprano, alto, tenor, and bass. But "four" is insufficiently magic for our assemblage here: "four" leaves out the dolphins and oilbirds and rhinoceros singers and the animals who just thrum.

So, all twenty quintillion of you, just go ahead and arrange yourselves however you want! As soon as there's more than one of you, you can be homogeneous or heterogeneous. You might sort yourselves by smelliness, sneeziness, spazziness, speckliness—speckled chachalacas can sort themselves from plain chachalacas, Holsteins from Jerseys. You can sort yourselves by biases and then again by sub-biases; there can be a reflective section and a section for those who are all reflex. There can be a section for the surreptitious—we're not sure who you are, but we noticed you arriving, obscured by the leafy branches, pampas grass, and toadstools you were carrying in front of you.

There can be an emergency section for the two- and three-year-old humans, who are forever losing their marbles, who act like the stars are sparkling them to death. We will use the emergency singers quite a bit in our program tonight, since most music could use a little emergency. With the toddler contingent, there will be no pathetic, droopy music, no songs of resignation. They may be joined by some emergency singers at the other end of life, too, the ones jonesing for time. Along with the emergency singers, there can be a section for emerging singers, like owlets, as well as submerging singers, like crocodiles.

There can be a section for those who feel like precursors, like all those people and animals in history. Precursors sing with a lot of presentiment. Or maybe you feel belated and sing with lots of remembrance, like you elderly koalas who remember your forest before it ceded to the suburbs. Or maybe you feel perfectly current, like the man of the minute: currency has a lot of currency these days. But however current you feel, remember that everyone here is as contemporary as everyone else, and as temporary.

If you are undiscovered, you are in good company, with millions of undiscovered species. The Tapanuli orangutan herself was an unknown till last year. Those of you who aren't sure whether you exist or not can sing with the Mongolian death worms. If you feel imperfect, you can join the likes of Abraham, Moses, and David, or you may find yourself gravitating to the perfect section, with the wind-up toys and the single-celled constituents of slime mold. If you do join the perfect section, your repertoire will necessarily be reduced, for perfection is only attainable in miniature. Anyway, music is a good form for the fallible, because mistakes made in music are like mistakes made in snow. Also, imperfections make someone a better wisher, and a better wisher is a better singer.

There can be a section for the thousand-songed singers, like the thrashers, and a section for the one-song singers, like the white-bellied go-away-bird—"*Go away, go away!*" Someone who can sing only one song is someone with a very stable identity, like an ice-cream truck. When an ice-cream truck joins the symphony, either everybody else has to play "Turkey in the Straw" over and over and over, or else the ice-cream truck has to stay quiet during the other pieces till finally everybody plays "Turkey in the Straw." With their one jingle, ice-cream trucks can evince one thing and one thing only—not death welling in sweet William, not a girl imploring her lover to remember her but not her wrongs. Though of course the more versatile musicians cannot dispense Choco Tacos.

We're sorry, but there will be no prizes awarded today, and if you came here hoping to sing about money, remember that money is a sore point with many animals. We couldn't really think of any indelible songs about dollars anyway. Oh, and another thing we should have said first off is that everyone must stop eating each other. One important foundation of music is that the musicians are not devouring, eviscerating, mutilating each other. Forgoing these pleasures, you may discover a different kind of pleasure. Spanish ribbed newts, please refrain from poking your ribs through your skin and poison-jabbing your neighbors. Tasseled wobbegongs, stop ambushing your little oceanmates. Humans, please turn your guns into kazoos.

That music is nonviolent is one reason we chose it; also because it transcends apathy, invective, and fatigue. Many of you must be so tired from all your trekking, to say nothing of your normal dam-building, web-spinning, burrow-digging, dish-doing. Many days are so exhausting that we conceive insipid, dishwatery philosophies. So let's have a song to spike our philosophies!—a solo, sweet as a julep, sung by a canyon wren. While she sings, we can think about how there is so much being in so many beings, but also so much being in one being. One little wren can fill a whole canyon with her silvery rallentandos. Tininess is no object for musicians, and neither is gravity. Are you a tiny musician? No worries! Are you a musician subject to gravity? No problem!

For even canyon wrens have bodies that are subject to gravity, but voices that are not, and this is equally true of guinea pigs. Although they look like lumps, guinea pigs sing not lumpish, leaden songs but whistly whirly-up songs. Sometimes we humans take this soaring property of the voice so far that our songs leave the world altogether, flying up to otherworlds. We're not sure if you animals have otherworlds or not, but it seems like there are plenty of songs to be sung about the marshes and grottos and simooms you have actually experienced, on the planet you have actually experienced.

If you sing from experience, it will be more interesting anyway, and experience is different from taxonomy, demography. Demography was always handing us scripts to recite and songs to sing, but every time we tried—dutifully we tried—we'd contract such violent fits of yawning that all we ever managed to emit were loud yawny yowly sounds. Many babies were born into the same social stratum on the same March morning on the same hospital floor in Orlando, Florida, but they do not therefore grow up to gaze at the same exact stars or think the same exact thoughts. More various, even, than the stars one might look at are the thoughts one might think.

Anyway, if we passed out songs for you to sing, we'd

probably get things wrong, like having the frogs sing about furballs. Even those who are "in the know"—who know that frogs don't get furballs but that they do contract a chytrid fungus that affects their ability to breathe through their skin—even *their* prescribed songs might not suit the frogs. If frogs are losing traction, if the world is becoming more frogless, then frogs are supposed to sing songs of desolation. But maybe they don't want to—even with all the troubles of our time, maybe it can still be fun to be a frog.

So take it away, everybody, sing your own songs! Sing the fungus blues, or pollywog variations. Sing of nursing your fourteen oinking infants, or scoring all those candy-cane beets by pulling them down into your tunnel underneath the garden, or losing your tree of life, or shivering all winter with your fellow bees to keep warm, or being caged for life with a peevish fellow hamster, or surfing the gnarly waves in Curio Bay with your fellow dolphins.

Of course, if there's one experience we all have in common—like the sun—it's that we all exist. Excepting (perhaps) the Mongolian death worms, everybody here squeaked onto the Existence Boat, while a bunch of other passengers missed it. Sorry, bub. Actually there are so many more bubs who *missed* the boat than bubs on board that, when we think about it, it seems anomalous to be alive. This is the thought that unsorts us from our infinite *divisi* back into an infinite *unison*—that all animals are anomalies, all ducks are odd ducks.

Now, all you anomalies go swim around a little, or rinse off in the rain and let the sunshine dry your wool or prickles or scales or snarls or plumes or togs, and relax and be ready in an hour to reconvene. Music can't be summoned—it is not a domesticated spirit, but wanton, like the wind. The wind is never going to hand you a diploma, but it might just blow your mortarboard away. We'll simply sing as well as we can, all night long, till everybody's eyes are sparkling. Even goofballs when they sing become sublimely lovely. Sing, you beautiful lumps, you beautiful buzzards and boobies, you beautiful galoots.

Sing, kiskadees, squeak, kinkajous, and laugh, kookaburras—run your gamuts! Everybody's like a banjo, everybody's got a gamut, a highest note and a lowest note and a range in between. Some gamuts might be smaller than others, but you can still do a lot with three or four notes. And with all of our gamuts combined—well, twenty quintillion is a number with more *zeros* than we can even count—whenever we'd hear the number "twenty quintillion" our heads would fall off. Combining all twenty quintillion of our gamuts will make for one infinitely unrunnable gamut. So after we've intoned, all night long, our various riffs and ditties under the various stars, then, just as our common star is coming up, we will gather all of our voices together and sing our finale.

The moths can begin with their soft songs, then the rhythm birds can join in, palm cockatoos knocking twigs on wood, prairie chickens booming. Then—*tutti!*—with juncos trilling, turkeys gobbling, leaf frogs sworping, eagles screaming, sea lions barking, babies bawling, elks bugling, and slugs—oh yes, throughout the whole piece will be interspersed the exquisite silence of the slugs.

Speckled and plain, perfect and imperfect, indigo-feathered, green-skinned, orange-toed, squashed of face, cracked of shell, miniature of heart, young as ducklings, old as hills, everybody raise your sweet and scrapey, bangy, twangy, sundry, snorty voices into such a song of amplitude as only we could sing, we waders and whittlers and melancholy woodpeckers, we Enzos and Ayahs and Wandas and Waynes, we hinnies and yaks and dingos and snakes, and all the little grebes indigenous to Earth.

ROBERT THE BRUCE GETS HITCHED

I have finally figured out who is buried in my backyard in Chicago. It is Robert the Bruce. I can tell by looking at the grapevines, for grapevines drink the blood of the body beneath them, expressing its spirit and revealing its identity. Our grapevines muscle over the fences and trample the factions of squash and scale the four-story building next door; and whoever was that indefatigable but Robert the Bruce? Sometimes the vines springing from the great Scot almost subordinate *me*, too—in the summer I have to bicycle through the yard at full pedal, outracing the tendrils grabbing at my ankles and spokes.

In fact, the whole country, writhing with mustang grapes and canyon grapes and riverbank grapes, seems sown with the bodies of zealots and swashbucklers and other fervent figures. They have no institution, those wild grapes; the sun is their vintner. Unschooled and unsupervised, they're like a self-raising child or a self-writing book. With all their oomph it seems they would even overwhelm fate—if fate were standing there in person, I bet those vigorous vines would whoosh over him and keep growing.

So our cup runneth over, but with wine that does not have a "soft velvet" flavor. If I were to crush and press those tiny blue grapes in my back-vineyard, I would call it Damn the Torpedoes wine—*Tastes of tenacity, pairs well with hay*. This is wine not for the moneyed lifestyle but for the pennied lifestyle; also for the turkey lifestyle and the skunk lifestyle.

Pinot Noir does not grow wild in my backyard. Pinot Noir is a finer, fussier, Frenchier grape that appears to issue from the remains of Little Miss Muffet—or, rather, *Mademoiselle Muffette*. Sweet as she was, Mademoiselle Muffette would not have wintered well in Chicago. She would have whimpered, frozen stiff, and fallen off her tuffet.

Now, thanks to the respective virtues of American and French grapes—durability and drinkability—there has been some missionary activity in both directions. In the nineteenth century, some American sand grapes sailed over to help their sophisticated cousins resist a root louse; more recently, a delegation of French grapes traveled to the Midwest to help the wild grapes become palatable.

Robert the Bruce and Little Miss Muffet are getting married and moving to Wisconsin, and their neighbors are beans and corn. "Take courage, Miss Muffet! Together we will be both tasty *and* hardy! You can show me how to be palatable, and I will show you how to survive a northland winter! Being brave is not just about storming the countryside in June. It also consists of *deep waiting*—waiting, through the long winter, for the warming star to angle back to us.

"In October, we will drop our leaves and go to sleep; we will squeeze the water out of our cells, so they don't explode when the killing frosts come; we will turn brown and dry, so we do not freeze inside. However fringed in icicles we become, however snowy our sleeves, we must not freeze on the inside, my dear Miss Muffet—not in our pith. Though the winter be long and harsh, we will refuse to give our hearts to ice." It is true: in the winter I don't have to sprint through my yard for fear of subjugation. I can walk right up to those sinewy climbers, for they are suspended—withered brown twigs encased in glassy layers of ice.

But winter is not the only problem for grapes like

Pinot Noir. The winter here is too wintry, but then the summer is too summery and the rain too rainy. Pinot Noir is not happy soggy; she suffers from bunch rot. Another problem is that the dirt is too dirty. For example, my tomato plants, as soon as I drop them into the ground, spring up taller than me and fall over and then leap up again and fall over again. Pinot Noir is used to growing at a more dignified rate, concentrating her powers in chalkier, gravellier, limestonier dirt.

If I had really wanted Pinot Noir to flourish in my yard, I should have deposited a ton of limestone in the ground and tilted my neighborhood up so we were on a drainable slope, facing east, for morning sun and afternoon shade. I should have had the Romans invade Chicago in the first century, planting *Vitis vinifera* vines to initiate their character development, and in the twelfth century I should have thought to import some Cistercian monks to jolly the grapevines along and keep their meticulous notes when they weren't praying. I, however, did not have the foresight of those Burgundians.

The wild grape's wisdom—don't take ice to heart—had struck me as true. But not even the truest true is unanswerably true. If you listen closely, a voice can always be heard in answer: "*Pardonnez-moi, monsieur,* but I am more than just palatable. I am a grape upon whom nothing is lost. If I am difficult to grow, low-yielding and easily distressed, it is because I am easily influenced. I do not damn the torpedoes, because I do not damn anything. You are thick-skinned, so everything is lost on you—frost, sog, monks, leafhoppers, torpedoes, tornados. But I am a thin-skinned grape—I feel everything—every fate is felt. I give my heart to ice because I give my heart to everything.

"The bodies beneath me in Burgundy are not those of frightenable figures but the skeletal remains of Jurassic creatures who swam around millions of years ago in the inland ocean. I register the inscrutably Jurassic, the inscrutably old—as well as the inscrutably new: in my berries are all my experiences inscribed and intensified. I taste of dews and droughts, of fearsome Februaries, of sunburn, and subtle, cooling winds from the north. I express things whose existence is only as yet proposed, by *les personnes*, and things whose existence has yet to be proposed.

"Things do not befall you, *monsieur*, you befall them, so your wine only ever tastes of you. Your wine is Vin de Moi—Wine of Me. During your one problem, winter, you suspend yourself, but I have many, many problems, and I am not good at suspending myself. Subject to danger, and chance, and change, I taste of whatever has befallen me. Yours is the wine of guarantee, the wine of certainty, mine the wine of chance—mine the wine of consciousness."

Thus replies the vine to the vine, as truly as he had spoken to her. The wine of certainty is very cheap, the wine of experience dreadfully dear. But oh to be impressionable *and* invincible! Wouldn't that be an ingenious combination—to take things to heart without being ruined, to be both sturdy and conscious, like bird and stone, or skin and bone, or like that double-minded, double-brave, dialectically composed Wisconsin wine.

HAUNTED BY HEDGEHOGS

How impoverished is a mind unhaunted by hedgehogs! Averse as I usually am to propaganda—most propaganda starts me lacing up my fastest shoes—I am entirely comfortable with hedgehog propaganda. Propagating hedgehogs in the mind, as well as newts and coots and camels and panthers and buffaloes and ouzels and bats, is a good cause, and it was the purported project of medieval bestiaries. The beasts were illustrated and described, and there was a small sermon on the meaning of each creature—"So it is with you, O man"—and then the animals entered your imagination, trumpeting loudly, teaching their young to jump across chasms, foraging for an herb called dittany.

Because they preceded the modern mania for veracity, these compendiums were rummy and reckless with the facts. Antelopes sawed down trees with their antlers, bears gave form to their pulpy cubs by licking them into shape, and panther breath was so sweet it enticed all the other animals to come close—except for dragons, who hate sweet breath. Hyenas had a special stone in their eyes, which, if you put it under your tongue, would enable you to see into the future. Some beasts were endorsed; others were not: "The whole of a monkey is disgraceful, yet their bottoms really are excessively disgraceful and horrible."

When an eagle's vision is growing dim, "he flies up to the height of heaven, even unto the circle of the sun; and there he singes his wings and at the same time evaporates the fog of his eyes, in a ray of the sun." A lion, "like the king he is, disdains to have a lot of different wives." After they are born, lion cubs are dead for three days, after which their father blows on them and they come to life. The lynx's urine crystallizes into a precious stone called Ligurius, but because of "a certain constitutional meanness" he scrapes sand over his urine so nobody can find it. Elephants live for three hundred years.

So the bestiaries are full of quaint misinformation and quaint composites, and quaintest of all is the animals' lifestyle: "They wander hither and thither, fancy free, and they go wherever they want to go." They are also full of quaint applications: the animals discussed are always parallels, symbols of something else. The hydrus, a determined-looking pink slithery fellow, dives down the crocodile's throat and then exits out of its stomach. It seems indecorous for both parties, but the crocodile is hell and the hydrus is Christ. The lion cub is also Christ, being resurrected after three days, and so is the rhinoceros, because he has one horn. Christ is very discernible, as are Adam and Eve and Satan.

However, the animal *itself* is less discernible. This is not animal propaganda after all: the crocodile is far less important than the interpretation of the crocodile. As if a crocodile were code for something else; as if somebody had a message to send and wanted to say it confidentially and so came up with this big scaly snappy greenish-gray submerged thing. If the crocodile is an encryption, then the crocodile needs deciphering—the crocodile is not really a crocodile, the dromedary is not really a dromedary, and the spoon worm is not really a spoon worm. They stand for something else, some message, made of words. Instead of words being symbols for things in the world, things in the world are symbols for words, and if the whole confusing pantomime were finally decoded, there would be no birds, or herds, or nerds: just words. The world deciphered would be like the ceiling of the Sistine Chapel with the figures sponged off and, in their place, cards reading "Angel," "Angel," "God," "Zerubbabel."

Caesar Augustus had a secret code in which he replaced a letter with the following letter of the alphabet, so "a" is "b" and "b" is "c," so the code word for "cookie" is "dppljf," and the code word for "nanny" is "obooz," and the code for "Snodjz Jzmrzr," that old opulent Arabian port, is "Topeka Kansas." Since the world as a code seems harder to break than Caesar's code, since cracking it seems more complicated than just shifting the alphabet one letter over on your little decoder ring, it can feel like you need somebody specialized to step in and translate the world for you, somebody like a pope.

Of course, there *are* decoder rings out there—theories advertised as being able to decipher the world—and since rival rings are sold by rival purveyors, rival popes, sometimes the decoding gets violent. Once, I was trying to populate my mind with beavers, visiting them at the zoo (where, after watching them sleep for hours one afternoon, I learned they are nocturnal) and in the Adirondacks, and watching them online. Underneath some footage of beaver kits squeaking in their lodge, the commenters had worked themselves into a fuming frenzy, each side claiming that the squeaky babies proved their respective theories about the origin of the world.

For the commenters' purposes, the beaver babies were nothing more than furry brown matter with which to clobber each other, and I imagined them all in a late-night rumble in the warehouse district, supplied with a wriggling pile of creatures, the combatants reaching for the nearest animal to brandish. It needn't be a beaver—it could be an armadillo or a bobcat or a little blue fairy penguin. You smash me with a penguin and I'll wham you with a marmot, you clonk me with a turtle and I'll flog you with an eel. The difference between combat and ideological combat is that with ideological combat any animal can be wielded, even unwieldy ones like whales.

So we use animals to prove our theories; so the application of the animal still takes precedence over the animal itself; so, as ever, we learn what we want to learn. If we want to learn that all our assumptions are corroborated by baby beavers, that is what we learn. If we want to learn that chickens *evolved* to be put into potpies, or that chickens were *created* to be put into potpies, then that is what we learn. Evolved chickens, created chickens, they're all meaty, and why would they be so meaty unless we were supposed to eat them? It seems like an animal could get lost between all our points, or at least compressed, but recently, early one morning (before they went to bed), I watched some beavers paddling around in a river near Jackson, Wyoming, and I noticed that they were not lost or compressed.

Anyway, I too have things I want to learn. From the otter shrew I'd learn how to be hard to trap. From ants I'd learn how to take out the trash regularly. From noisy night monkeys I would learn to stop being such a square, and from the whale I'd learn to be unwieldy. From pipits I would learn to imitate pipits, from owls how to imitate owls, and from marsh warblers I'd learn to imitate every bird I've ever heard. From pangolins I would learn how to imitate pinecones, from goats how to be happy in company, from pandas how to be happy alone, and from the ovenbird I'd learn how to be incognito, overlooked—how to be secretly present.

From medieval animals I would learn how to wander as hither and thither as I can, how to hold off the dragons with my sweet breath, and how to outlive the wisdom about me. From the modern beaver I'd learn to paddle around in the cold Wyoming water, plump and hardworking and nobody's spokesbeaver. Modern chickens are not very fancy-free but they are still very chicken: from them and from all the animals I would learn how to make do, make do, and not be anybody I don't have to be.

BEANSTAN FOR PET

One summer afternoon, I was reading poems by someone who had evidently soured on the world—and not just on one sector of it, like planetariums or dolphinariums, but on the entirety. As I read, I was feeling more and more convinced: *Too true*, I thought, *the world's a bust, please pass the arsenic.* But then I looked out the window and saw the trees in my backyard dancing to what seemed to be zydeco music. The real, rancid world was obscured by those shimmery, shimmying trees. I could not see the world for the trees.

That was the afternoon I got snookered out of my philosophy by aspen trees, as I have on other occasions been snookered out of my philosophy by sappy little birds and countless times by one little dog, one fluffy little orange dog capable of snookering anyone out of any philosophy. It would be much easier to maintain a clear view of the world without all those obscuring trees and birds and Pomeranians. Of course, if you want to see the world unobscured by Pomeranians, you can go to Yellowstone, where they are allowed only in parking lots or handbags. Nowadays Pomeranians are bag dogs, but they started out as bog dogs, protecting the bog people from the bog wolves.

I had a Pomeranian and a Labrador—a little bad dog and a good big dog. The big dog had thought for many years and was in possession of great understanding. The little one attacked anything he did not understand, and he did not understand anything. The Department of Defense he headed up was extra-vigilant and extra-berserk; he was all valor and no discretion. With no bog fiends from which to protect me, he protected me from all the nice neighbors, especially the nice neighbor lady in the hot red shoes. Sometimes he mistook *me* for a nice neighbor lady and savaged my ankles.

If most Pomeranians seem like toys you'd find at a fancy toy store, posh and well crafted and safe, Beanstan would have been a broken toy in the trash can out behind the toy store, all jagged edges and faulty wiring, headed to the city dump unless fate had some other crazy idea. Ralph Waldo Emerson wrote, "Your goodness must have some edge to it—else it is none." Beanstan's goodness had so much edge to it that Emerson might have said to him, *Your edge must have some goodness to it.*

We tried to program some goodness into Beanstan but found him nearly unprogrammable, nothing like software. Some of our programming did turn into superstition, however. When he was thirsty he would twirl by the sink: because he had learned to twirl for treats, twirling should also turn the faucet on, open the door, make the chunk of cheese on the counter float down to the floor for him. Until the day he died, Beanstan believed that twirling would make his wishes come true.

Sometimes when I came home from work he got so excited he bit me, and sometimes he found a sock to bite instead. He lost his mind with joy, he lost his mind with fear; he had a very loseable mind. People who slept in the same bed with him said to each other, "Don't let the bed-dogs bite." When guests stayed at the house, we'd have a special activity called "Feed the Wild Animal," in which Beanstan's leash was tied to the oven door and they could stand across the kitchen and toss corn chips to him. One winter, two people who had hit bottom came to live with us, and sometimes, if Beanstan had not been properly secured, we would come home to find them standing on the kitchen table or trapped in the bathroom. (When you hit bottom, Beanstan will be there to bite you.)

Perhaps he was born into the hollow of a tree, and then the tree was struck by lightning; or perhaps, just as he was trotting by the open door of a storefront church one summer evening, they were performing an exorcism. I was not supposed to fall for any of my foster puppies, and I did not fall for so many of them—Pepsi, Alfie, Bailey, Wrigley, Beanstan's sister Penny. Penny was equally Pomeranian, but I let her go without mounting a fervent campaign, without making signs and posters reading "Penny for Pet."

My "Beanstan for Pet" campaign was a success, except instead of becoming my pet he became my dread fluffy little master. If Beanstan had been the Pomeranian in Chekhov's story "The Lady with the Dog," the title of the story would have been "The Dog with the Lady" or "The Lady Chasing the Dog Chasing the Decent Person," and there would have been no romance. One winter morning, Beanstan and I were walking around our Chicago neighborhood when a woman crossing the street fell down on the ice, and he tried to bite her. He would take anybody on, persons sprawled on the ice, mastiffs, drive-thru cashiers, ice-cream trucks, everybody except for the inflatable golfing snowman who came to golf in the neighbors' front yard every December.

Beanstan didn't necessarily need a *something* to chase; sometimes he'd accelerate around the backyard by himself, petticoats swinging wildly, chasing after nothing (or maybe it was the Holy Spirit—it is hard to tell the difference between the Holy Spirit and nothing). Once, in an Illinois forest, he hurtled away and didn't come back until an hour later, trembling with pride and ecstasy, his silky fur completely lacquered in green excrement. He did appreciate the finer things: he loved to wake up at 5:00 a.m. to be let out into the yard to lick the grass. As you connoisseurs of dew know, there is no dew like five-in-the-morning dew.

He would swim for his life, but for no other reason. When the big dog paddled after tennis balls in Lake Michigan, he would stand at the edge of the water, shrieking and twirling. Once, we took him floating down a warm river, and he swam crosswise to the flow the whole time, desperately liking the looks of the shore. When we moved to the mountains, we hoped that Beanstan might find himself. As it turned out, he did not find himself here, but he did find magpies and moles and moose and dead moose, and he learned to leap high because of all the snow. As Denmark teaches people to drink deep, so Montana teaches dogs to leap high.

Most days, he hung out under the piano, a little wild animal on a red cloth bed. Many evenings, he and the Labrador constituted the audience for private piano concerts. Annabelle would stay for most of the performance; she liked "Moon River" and "Alice in Wonderland" and "What'll I Do." If you listen to the words of "Moon River" or "What'll I Do," you might notice that the song is not about a group, a class, a breed. Nobody would write a song like that about a taxon, only about a you. Sometimes even now when I see an orange Pomeranian with panache, my heart starts to pound, but then I get closer and see how peaceable it is. All those Pomeranians I don't know are like all those people I don't know—only from far away do they make my heart hammer.

Both audience members under the piano enjoyed the Irish waltzes and the Scriabin piece and Chopin's preludes. "Prelude" is a funny name for Chopin's preludes, because they have no movement following them. For many people, the days are like Couperin's preludes—normal preludes, minor warm-ups to mainer movements. Monday's a prelude to Tuesday, and Tuesday to Wednesday, and the next day, until it arrives, is always the important one. But for dogs, the days are more like

Chopin's preludes—preludes to nothing, similar to how a dog can chase around the yard after nothing.

Some hymns always run amok, who knows why. Annabelle was no fuddy-duddy; she liked "Blue Bossa" and the tango and "Wade in the Water," and everybody else (Beanstan) seemed to enjoy those songs, too. But when "Marching to Zion" began, Annabelle would always retreat to the remotest bedroom. Obviously Beanstan was not the kind of concertgoer who would stick around just to be polite. When anyone whistled, he tried to paw his ears off. But there he'd lie, as the hymn fell apart and the pianist's feet stomped and pencils and papers fell off the rumbling piano, and instead of marching it was more like slamming and lurching to Zion, and he would close his eyes and drink it in, as if bedlam wasn't antithetical to his soul, as if chaos were his clover.

THE LAND OF IMPORTUNITY

When introducing two parties, it is important to keep their social standing in mind. First you are to address the one with more status—as in, "Successful Actress, I'd like you to meet Usherette," or "Monkey-Faced Bat, I'm pleased to introduce you to Mongoose." Monkey-faced bats are endemic to Fiji and thus have seniority over the invasive mongooses. Escaped plants rarely give you a chance to make a formal introduction. Ornamental plants with modest rainfall needs escape their little patch and rush around introducing themselves—"Hey, I'm Periwinkle!" or "My name is Butterfly Bush, what's up?"

In 1691, there was an escaped Frenchman named François Leguat who introduced himself to Rodrigues

Island. He had a home in France and "Persons that were dear to me" and was not looking to leave his home or his persons. But the mind that sallies out may oblige the body to sally out as well. Having read the Bible not in the supernatural language (Latin) but in French, he had started to think illegal thoughts. So he sailed away in a frigate with some Jeans and some Jacqueses and got marooned on a little island in the Indian Ocean for two years. "I was as it were forc'd to give way to Importunity."

Importunity, the opposite of opportunity, is inconvenience, a common experience for exiles. Some French Protestants sailed to Brazil, where they were eliminated for not converting. Some sank before they could be eliminated in Florida. (Mortality was high in those times, but so was immortality.) Some stayed in France and were eliminated via conversion. Some fled to forests and lived off boiled pinecones, perhaps, and others fled to Russia, where they made descendants who made descendants who made descendants who made Fabergé eggs.

Whether your thoughts are legal or illegal, it can be good to get out of your milieu sometimes, maybe even for several generations. For invective is infectious, and before you know it, it's spread from your pamphlets to the rest of your life, and you are preparing an invective pot of porridge, playing an invective piano, casting an invective shadow. But open up that red enamel egg, and inside you will find not invective but a pale-yellow rosebud. Twist a pearl on the Lilies of the Valley Egg and out pop three tiny portraits. Out of the Bay Tree Egg flaps a little green bird, singing a song. Other eggs contain other delightful surprises, none of which is invective. Invective is rarely a delightful surprise, or a surprise at all.

Some Huguenots fled to South Africa and Wales and New Jersey, where they made wines and textiles and little Huguenots and interesting observations. But what is a Huguenot to do with his observation where there are "neither Cities nor Temples, nor Palaces, nor Cabinets of Rarities, nor Antique Monuments, nor Academies, nor Libraries, nor People" on which to employ it? On Rodrigues Island, where Leguat inadvertently lived with turtles for two years, there were no temples, no libraries, and little conversation. Who is terser than a turtle?

But any material can be material for observation. Especially diverting, after you've been at sea for a while, is anything other than water. Leguat observed melancholy birds who hung their heads into the sea, and fish flying out of the water "to avoid the Persecution of the Goldfish." He saw Saint Elmo's fire curl around the mast, and there was music: his ship was followed by four singing swallows, and once, while sleeping on the sand of an island along the way, Leguat and his companions were awakened "by the braying Musick of a Rustic Regiment of Asses." It is not escapist to lie on the beach and enjoy the donkey music while you are escaping from the converters.

And he did observe rarities, though not in cabinets. There were rare stars, rare sea cows, rare reds and greens and blues—on Rodrigues Island, where the ship dropped the men off, he observed lively-colored blue and green and red and black lizards, and tortoises with no qualms, and bats as big as hens, and birds with the wings of a bird who never plans to leave. Animals on a remote island are not very progressive; many of them have made their peace with gravity. Nor are they very subversive, since the Bible has never been translated into Gecko or Parakeet. (Though there *were* little birds who would fly up behind the men and snatch their bonnets: "These birds made a pleasant War upon us, or rather upon our Bonnets; they often came behind us, and caught 'em off our Heads before we were aware of it.")

For the Protestants who landed there, it must have been strange that the humming of hymns was no longer a subversive activity. What's a Protestant without his subversive humming? What's a Protestant with nothing to protest? Maybe just a person with vegetables to hope for—"We had great hopes of our Artichokes"—and trees to climb. One fellow "us'd to perch upon a Tree in an Inundation," and another was "always singing of Psalms." They bathed in the sea, surrounded by cordial sharks. They made liquor out of palm trees, as well as hats and umbrellas. As for Leguat: "Know, kind Reader, that my chief Employment in this Desart Island was thinking."

His chief employment was thinking: he thought about "Terror and Custom, two terrible Tyrants." He thought about air and precipitation and animals hatching out of eggs. Out of every egg came a surprise—not enamel animals but parakeets and pigeons and fodies and little giants. There were two species of giant on the island—giant tortoises with long necks, who ate the tall plants, and giant tortoises with short necks, who ate the low plants—and both kinds were friendly to the humans. Understandably, tortoises would be curious about such a souped-up animal.

The giant tortoises herded together by the thousands, like slow green benign mobocracies. There were solitary animals on the island, too, birds called solitaires, who were "seldom seen in Company." Like dodos, they were pedestrians, but just because they didn't fly didn't mean they had no flair. From the big knobs on their small wings and the fractures in their skeletons, it seems they had a flair for combat: solitaires were violently solitary.

Their solitude started in the nest, where each had been the only egg. Their parents sat on them in the forest and fended off any solitaires who wandered by. Of course, a good forest is an important part of the solitary lifestyle, and Leguat found the island of Rodrigues well forested. Its little mountains were covered with trees: "Their spreading and tufty Tops, which are almost all of an equal height, joyn together like so many Canopy's and Umbrello's, and jointly make a Cieling of an eternal Verdure." (Trees make decent foresters if no one else is available.) (A tree also makes a nice umbrello.)

Parents on a remote island can get away with putting all their children in one egg. Except, once the pigs arrive, the one-egg plan is unfortunate. In the eighteenth century, cats and pigs and Dutch people sailed to the island, and the solitaire, who was "delightfully edible," lost its solitude. After it went extinct, it glimmered for a while in the sky: in 1776, somebody turned the Hydra's tail into a solitaire, but then in 1822 someone else turned the solitaire into Noctua the Owl. (They were always rejiggering the stars back then. It's not like somebody nowadays would be allowed to change the Scorpion into a Raisin Bun.)

Anyway, the artichokes failed, and meanwhile the younger exiles were getting tired of the lull in their lives. "They wou'd have Wives. *Wives!* said they, the only Joy of Man." Leguat was older and knew that wives were not the only joy of man. But, just as importunity had brought him there, so importunity made him leave: Leguat did not wish to stay behind all by himself. The men built themselves a bark and wrecked it and built another bark. The sea supplied them with one massive oak beam for their boat, and necessity supplied everything else—out of need, the non-carpenters became carpenters and the non-rope-makers became rope-makers.

Before they drifted off to Mauritius, Leguat composed a poem for the island, thanking it for its palm wine, its melons, its clear water, and the "precious Treasure of Liberty." He prayed that astrologers would never defile the island, or persons who write inferior poetry or "mad pedants" or "presumptuous Earthworms" who try to explain mysteries.

As it happened, most of the mysteries on Rodrigues Island were eliminated before they could be explained. They were too innocent, too friendly, too pedestrian. The pedestrian mysteries who had forgotten how to fly got eaten, as did the radiant little mysteries who came down from their trees to nibble melon out of human hands. The giant mysteries lumbering along in genial conglomerations found their plants going up in flames, to clear rectangles for agriculture. There was a bright-green bird on Rodrigues called *Necropsittacus*, which means "dead parrot." Along with the solitaire, the giant tortoise, the starling, the parakeet, the gecko, and the pigeon, the dead parrot went extinct.

If only mystery could go into exile instead of going extinct. If only the mysteries of Rodrigues could have found a bark and floated out to sea—the tortoises content to jostle together, the solitaires fending each other off—and finally washed up on an island where grew trees like this: "There's a wonderful fine Tree at Rodrigo, whose Branches are so round, and so thick, 'tis impossible for the Sun-Beams to penetrate thro' it: Some of these Trees are so big, that two or three hundred People may stand under them, and be shelter'd from the Sun or the Weather." A tree like a pavilion—an island pavilioned with trees like that, with leaves like soft green heart-shaped hands—could have protected, from the presumptuous, so many multitudes of mysteries.

IN LIEU OF A WALRUS

They're always talking about being up a creek without a paddle, as if a paddle were the only thing you could be up a creek without. Of course, when you're up a creek the paddle is the *relevant* thing, but you can also be up a creek without a parakeet or a pagoda or a Greek-Syriac lexicon. I've been up a creek without a tuxedo before, and once I even found myself up a creek with no tuba.

Similarly, when they call someone ill-informed, it seems like they're always referring to relevant issues. But someone can be ill-informed about irrelevant issues, too. Not only do I not know much about the Supreme Court—I also don't know much about the Grindletonians, the glyptodonts, the Quasi-War, the Dunker denomination, or "Irish Marvels Which Have Miraculous Origins." Nor do I know much about Irish marvels with ordinary origins. Not only can I not bake sandwich bread, I also cannot bake knishes or kringles or biz-cochitos. Not only do I not have millions of dollars—I also don't have millions of kopecks or drachmas or shekels or millions of those big stone disks they use on the island of Yap.

Everybody wants to be relevant these days, and everybody wants everybody else to be relevant, too. It can seem that relevance is the only virtue, or that among virtues relevance holds a despotic ascendancy. But of course relevance is relative. What is relevant to an English schoolgirl—impressive vocabulary, schoolroom society, simple rules to avoid getting poisoned, being smarter than Mabel—may not be relevant to a walrus or a crab or an insane duchess.

If you find yourself oppressively preoccupied with relevant issues, just start a conversation with an eaglet and a duck; go croqueting with a flamingo as your mallet; attend a tea party with a dormouse and a hare and a hatter. This will make a hash of your relevance, for it will not even register with your interlocutors.

I realize I am making it sound too easy; I know flamingos can be suspicious and eaglets are not always in the mood for conversation. I also know that walruses often have prior engagements—how many times have I sent out the invitations, set up the card table out by the poppies in the backyard, prepared the tea and buttered the toast, but then only people show up, all of them sane? Being snubbed by walruses—this is why I read, because some writers are almost as good as walruses for joggling the head, freeing it of relevant concerns. Below is a list of the writers I turn to whenever I cannot find a walrus.

HAFIZ Presumably there were current issues in fourteenth-century Persia. A writer could have been relevant even then. However, Hafiz wrote poems about swapping jokes with the sun, the universe being a tambourine, rabbits playing cymbals, planets gone crazy. Because of his absurd topics, Hafiz is now as irrelevant as ever. Hafiz is timelessly irrelevant. Here is part of a poem that appears to describe his own writing process:

The mule I sit on while I recite
Starts off in one direction
But then gets drunk

And lost in
Heaven.

THE ANONYMOUS AUTHOR OF *THE CLOUD OF UNKNOWING* First of all, is there anything more irrelevant than anonymity? Furthermore, this mystical fourteenth-century Englishperson directs you, the reader, to pursue the *absence of knowledge*, to pitch all of your information into the Cloud of Forgetting and to pitch *yourself* into the Cloud of Unknowing and make your home there. If we all took Anonymous's advice and relocated to the clouds, I wonder what would become of the phone factories and the fact factories.

OVID A charming girl turns into a charming cow. Her miserably immortal father laments: "For me death cannot end my woes. Sad bane to be a god!" All the daughter can do is moo. Someone turns into a mint plant, someone turns into a screech owl, someone turns into a constellation. I wonder how attached they had been to their previous identities.

Imagine the person who for twenty years has been establishing her professional identity as a person, publishing pro-person articles in academic journals, participating in person panels, etc., but now, right after having scored an interview for a plum university job as Professor of Personness, she is turned into a cucumber plant. Say she decides to go ahead and interview anyway: she takes the hotel elevator up to the fifth floor, taps her green fruits tentatively on the door to the room where the hiring committee has convened, shuffles in, takes her coat off, and the committee says, "Oh, okay . . . Well, um . . . we . . . wow . . . from your CV we had you pegged as a person."

GOD That old author of the Fourth Commandment would give the animal workers (and servants and foreigners) every seventh day off. If followed, this whimsical commandment would slow down the dollarization of the world.

JOHN MILTON Because pamphlets are so flimsily made, they tend to be written about urgent, pressing, timely topics, like peer pressure and pickles and salvation. But of all the pamphlets I have read, pamphlets regarding *New and Old Pickle Recipes* and *How to Plan Your Children* and *How to Repair Zippers* and *How to Go to Heaven* (a clear behavioral plan that does *not* include the reciting of poems while sitting on a sozzled mule), my favorite is an old obsolete one called *Areopagitica*, in which Mr. Milton tells the church to butt out. This was back when, before a book was published, it had to first be approved by a bunch of interfering friars.

To fill up the measure of encroachment, their last invention was to ordain that no Book, pamphlet, or paper should be Printed (as if St *Peter* had bequeath'd them the keys of the Presse also out of Paradise) unlesse it were approv'd and licenc't under the hands of 2 or 3 glutton Friers. For example:

> Let the Chancellor *Cini* be pleas'd to see if in this present work be contain'd ought that may withstand the Printing.
>
> *Vincent Rabbatta, Vicar of Florence.*

> I have seen this present work, and finde nothing athwart the Catholick faith and good manners: in witness whereof I have given, &c.
>
> *Nicolò Cini, Chancellor of Florence.*

> Attending the precedent relation, it is allow'd that this present work of *Davanzati* may be printed.
>
> *Vincent Rabbatta, &c.*

> It may be printed, *July* 15.
>
> *Friar Simon Mompei d'Amelia,*
> *Chancellor of the holy office in Florence.*

Of course there are no glutton friars signing off on our books anymore. The most gluttonous among us—the grizzly bears—are so busy being gluttonous they don't have time to scrutinize potentially subversive material. This is why there are no craven books beholden to bears.

In some modern societies there are no censors whatsoever, except for the most censorious censors of all—the censors in the mind, those officers policing our thoughts to ensure there is "nothing athwart." Nothing athwart convention, nothing athwart consensus—nothing athwart *relevance.* Our thoughts could be starbound; why do we think like personnel? At what point do persons become personnel? They sure as pandemonium don't start out that way. In any case, as fat as those old friars were, they weren't as fat as God, who does not censor us at all.

NON SEQUITURS

With 7,839,056,589 members alive, the human species is obviously not endangered. Our fins are not wanted for soup; our floating islands do not intermittently sink to the bottom of the lake; we are not like the pygmy sloths, with fewer than a hundred of us left, hanging upside down from our mangrove branches. Nevertheless, humans *are* endangered *individually*, like that thin-shinned old auntie eating her potato supper in Carefree, Arizona. In Carefree a cactus could fall on you at any time.

Once, there lived, in the land of Uz, an especially endangered man named Job. First his sheep got burned up, then his camels were carried off by the Chaldeans (who must have had strong arms). Then his ten children were crushed by a collapsing house, and after that he developed painful sores. Then three of his friends, who must have been swallowed by some category, some ideology that caused them to think exactly alike, came to rail at him on his ash heap. Talking to one of them was like talking to any of them: Zophar might be Eliphaz, Bildad might be Zophar, Eliphaz might be Bildad, Zophar might be Bildad, and so on. Zobileliphardadphaz's theory was that good people do not suffer—therefore anyone who suffers must have done something wrong. This theory, like all theories, is able to absorb miniature experiences—mini-woes, toe woes, yo-yo woes, getting plinked on the head by a Ping-Pong ball. For the life made up of no-biggies, any theory will do.

Job's experience, on the other hand, was not the absorbable kind. Job is always getting called patient, but really he was only patient for two chapters. The following thirty chapters recount a ferocious conversation between Job and his friends—between Experience and Theory. Theory tries to muffle Experience, Experience gags on the Theory. Theory says, "Abominable and filthy is man," and Experience says, "God has wronged me." Theory says, "You subvert piety," and Experience says, "I am not inferior to you."

Anyhow, if somebody has a head you can always look

over it. Job says to his friends, "All of you are quacks," and starts directing his complaints up to the one who had taken him by the neck and shaken him to pieces. "I insist on arguing with God." Maybe you have been part of a band that felt like two-thirds of a band, or part of a conversation that felt like two-thirds of a conversation, the absent person being a dream drummer or someone with all the answers. The funny thing about the conversation between Job and Bilzophazeliphardad is that the missing person actually shows up—except, instead of answers, he brings lions and lightning and various other non sequiturs, like donkeys.

1. "Who hath sent out the wild ass free? or who hath loosed the bands of the wild ass?" The speaker is apparently referring to himself—*That's right, it was me, I freed the donkeys!* Intervening in the serious intellectual-theological debate is: the person who let the donkeys out.

Now, I have read books about how to find an escaped donkey—first you are supposed to think about where you would go if you were a donkey. But I have never read a book on how to set a donkey free. The fellow who frees the donkeys is probably the same fellow who doesn't farm the chickens and doesn't geld the bulls.

2. Additionally, he makes behemoths. "Behold now behemoth, which I made with thee." Typical of megafauna, the behemoth carries with her a standard that dwarfs other standards. You might have high standards for your household and run a tight ship, but as soon as a behemoth runs through the front door—tight ship, loose ship, who cares? When you are beholding a behemoth, your standards suddenly seem spurious, as does your identity. Ostensible identities—like being female, fashionable, sectarian, second banana, etc.—are interesting to think about but only if one is mixing with minifauna.

3. He spends a whole chapter talking about the megafauna of all megafauna—the leviathan, who "maketh the deep to boil like a pot." He does not mention pygmy sloths; pygmy sloths are too little and placid to make the deep boil like a pot. If they drop down from their mangrove branches into the seawater, they just paddle peacefully over to another tree. Pygmy sloths might not scramble your identity, as do behemoths and leviathans; still, it's always good to spend time with somebody who is more upside down than you.

4. Eccentrically, when he discusses the ostrich, the speaker refers to himself in the third person—"God

hath deprived her of wisdom." But even though the ostrich is a ding-a-ling, she can run very fast. You don't have to be wise to be fast. Nor do you have to be wise to be a megafauna.

5. "Wilt thou hunt the prey for the lion?" "Doth the hawk fly by thy wisdom?" God seems to get a bang out of how autonomous the animals are and how they don't need our help. This can be hard for us to take, since we really love to help. Someone's probably out there trying to be a flying coach for the hawks or a life coach for the ostriches, and someone's probably planning a prey-hunting conference for the lions, imagining whole prides traveling out to stand in line in the hotel lobby, waiting to receive their name tags and conference packets, perusing the schedule so as to attend the most professionally profitable panels.

6. "Can you bind the chains of the Pleiades? Can you loosen Orion's belt?" The constellations used to be more bindable and loosenable—just generally more changeable. Orion was pursuing the seven sisters when they got changed into stars. To me it looks like Orion was changed into stars too. But anyway, no one is meddling with the sisters anymore or taking Orion's pants off.

Now, Elibilipharzophaz, in his homilies, had made it sound like God was unimpressed by moons and stars: “Even the moon is not bright, and the stars are not pure in His sight.” But when God gets to speak for himself, he sounds pretty starstruck. Perhaps the friends—like most people who speak for God—were not really speaking for God but for themselves, disgusted by moons and stars and men. If man’s a maggot you don’t need to cry for him.

7. Among all these enormous non sequiturs—lions, leviathans, stars—is an itty-bitty non sequitur, a non sequitur to the non sequiturs—a tender little shoot. Truly, what is more of a non sequitur than spring? In winter, you can spend your days in the counting room, where figures follow from figures, but go out in May and you will see pale-purple pasqueflowers following from nothing. As destabilizing as it would be to be resurrected, that is how destabilizing spring is. It is also destabilizing to be born. Look at all those destabilized babies out there. Life is the great non sequitur.

8. “Who put wisdom in the hidden parts? Who gave understanding to the mind? Who is wise enough to give an account of the heavens?” God speaks mostly in questions, while Phazbileliphardadzo speaks mostly in answers. You

would think it would be the other way around—the humans with all the questions and the deity with all the answers. Supposedly, having the answers gives one authority, but often answers seem vulnerable, like sheep, needing a lot of protection. And as the shepherd needs her sheep, so the dogmatist needs his answers. How depressing to be a shepherd without sheep or a dogmatist without answers.

9. "Hath the rain a father?" Imagine trying to be the rain's father: *Stop that, Rain—no, stop, Rain, no, no, no, come this way, Rain, no, I told you to go to* Sappington, *not Silver Star again!* Rain has its own priorities, which seems to be the main thing that all these phenomena have in common. Peacocks, hail, hawks, hell, even some humans have their own priorities. Salted away behind the loudest people in the world are about 7,818,300,000 others, some of whom maintain their own priorities. You might see some of them at the DMV or the grocery store or the library; if you really want to see some humans with their own priorities, look for the littlest library-patrons, the ones much littler than senators.

10. "Canst thou send lightnings, that they may go, and say unto thee, Here we are?" When you call out "Yoohoo!" do all the little lightnings come running? No

they do not, lightning does not listen to you, nor does it distinguish between patriarch and heresiarch, pumpkin and bumpkin, giraffe and giraffe. There used to be a convention that lightning traveled only downward, but then we discovered pink electrical sprites, which flew in the face of that convention. That is what happens when you have a convention: things fly in its face.

And the creator goes on about snow and hoary frost and hungry baby ravens and hungry baby lions and hungry baby goats. This had been a serious conversation till God showed up. One would have thought that he'd be more religious, that he'd talk about intangible things, that his thinking wouldn't be so heavily influenced by animals, that he'd at least give his listeners some kind of takeaway. However, trying to find a takeaway from God's speech feels like trying to find a takeaway from a donkey. The donkey being every inch a donkey, any takeaway has to be Scotch-taped to her withers, and any takeaway you have to tape on to something is hardly a takeaway. In any case, with his leviathans sneezing lightning, God does manage to end the many-chaptered dispute: the dogmatists recede and God gives Job more sheep and camels and children and no one dies, especially the children.

But that was then and this is now, and the distance

between now and then is always growing. Here we are, getting nower and nower, while the book of Job is getting thenner and thenner. It's hard to imagine anybody naming a daughter "A Flask of Blue Eyeshadow" anymore, or being consoled by receiving a shipment of six thousand camels. But still it hails, still it rains; the clouds are still ungovernable, the baby animals are still hungry, and some of our conversations are still in need of intervention. If God does not step in with his arms full of baby goats and ditzy ostriches, perhaps we could try his rhetorical method ourselves.

When a conversation starts to feel like a dead horse that you and your friends have been hauling around for months, from campus to your apartment and back, down three flights of stairs, along seven blocks of sidewalk, and into a crowded diner, and after that into a moody little bar, and the horse is only getting deader and deader, so that by two in the morning you're so sick of the dragging and the stench that you try to bury it in the park, but the hole isn't quite deep enough, so horse hooves are sticking up out of the ground and you will surely get a citation and be made to dig it up: if it feels like you will never be free of such a tedious, putrid burden, you *could* try introducing a live horse into the conversation, especially a laughing one: "He saith among the trumpets, 'Ha, ha.'"

And if someone is trying to get you to submit to cautious pious religious clichés, to betray your own experience, to retract and recant and retreat: instead of saying uncle, you could always say, "I am not inferior to you," and go fetch a leviathan. If you can't find a leviathan, a goony bird will do, or a linnet. Linnets have no tenets; any animal, in response to religious dogma, says, "That's just religion talking." Dogness defies dogma.

Of course religion is not the only dogma, animals not the only confounders thereof. In response to some explainer of the brain trying to make you defer to images of little brain-blinks, you could say, "Excuse me for a moment," and run get a string quartet to play Beethoven's String Quartet No. 15. Composed after great suffering, after Beethoven had lost his hearing, this quartet he called his "Holy song of thanks to the divinity, from one made well." If you can't find a string quartet, you could ask the Texas Tornados to sing "Hey Baby Qué Pasó?" For, in reply to those blippy little blinks, almost any music says, "That's just neuroscience talking."

Other dogmas pretending to be in league with the universe would commandeer experience, crush it with words. But experience in extremis is not to be commandeered. Experience in her extremity is proof against command. "I will maintain mine own ways," says Job, and at the end of God's animal-mad speech, he says Job was right and the friends were wrong. They weren't fogeys in league with the universe, just fogeys in league with other fogeys. Phazipharibilizo, you lie: It's not true that the moon's not bright, that the stars aren't pure; it's not true that man's a maggot. It is extremely not true that the innocent do not suffer. The stars are pure and the pure may suffer terrifying things, burning terrifying bright, like the well-made Beethoven, the well-made Araminta Ross, the well-made man named Job. The mortal is inferior to no one.

PEDESTRIANS

It takes so long to learn anything truly important, like how to play the harpsichord. First you have to acquire one, then heave it upstairs into the attic and muffle it in blankets, then practice night after night while your family slumbers. After many nights of muffled practice comes "Frère Jacques"; then some erratic arpeggios; and long after that comes Handel and the consequent euphoria. To play euphorically on the harpsichord or trombone or that otherwise tedious instrument, the mind, is a lifelong undertaking. But of course lifelong is not that long, which is why we have to start learning as soon as possible, boa babies learning how to constrict, baby Clydesdales learning how to be mastered, baby mustangs learning how to be masterless, titmouse chicks learning to tink and squink and zeedle.

Upon emerging from the egg, barnacle goslings will learn that the heat they had been swiveling toward has a source—a mother and father. They will learn that the dimly apprehended glow has a source, too—a sun. In the long run, they will learn about the mossy greens, sedgy greens, rosemary greens, and other greens of Greenland; that saxifrage tastes bitter at first and then turns sweet; that flying runs in the family. With time, they will grow strong pinions to replace the silky baby fluff and will fly great lengths to winter in the Hebrides; they will fly over water, over stands of trees, over the Standing Stones of Callanish, over the heads of that managerial animal, man. For their first seven weeks, however, they are fuzzy pedestrians.

So far, so realistic. But then there is a wrinkle in the realism. A few days after hatching, weeks before they learn to fly, they have to fly, off the four-hundred-foot precipice where they are nested. Their parents cannot carry them down and cannot be bringing up little bits of grass from the valley all day long for weeks. The parents fly away, and the babies must follow. It is like firefighters having to fight fires three days after they are born. Baby firefighters and exegetes and mariachis—they all get some years to hone their abilities before they perform; even Peter Pears did not immediately command the stage at Wigmore Hall. Imagine baby Sir Peter Pears singing, *Citizens, I will deliver you!*

Of course, back when Greenland was grovy, birds could build their nests in magnolia trees. After the babies had hatched and their winglets had dried, they could tumble down and start browsing, and the tumbledown was comical. But then the forests froze and fell over, and without a forest it is hard to keep a secret. Some birds decided to take their secrets to rocky islands off the coasts—a choice that demonstrates cunning and craft. Some birds chose a less crafty, more vertiginous site for their secrets. Goslings can fly only slightly better than ponies, but still they must get down off the cliff, and no one is at the bottom holding a poncho.

Then they will learn that life was ducky in the egg. Their shell had been strong enough to seat a grown-up goose, commodious enough to house somebody passively assembling, constraining enough to clear them of all accountability. Good eggs or bad eggs (some eggs are exploders), all eggs are good determinists, able to trace back to antecedents every condition of their lives. Eggs are good determinists like pinballs and shadows and the tin eggs of a wind-up goose—a determinist herself, able to attribute her every action to how many times the key was turned, in which direction she was aimed, and how many eggs were poked up her bottom. But once her eggs are laid, the lineage stops: little tin determinists never hatch out of tin eggs. Determinism peters out.

Nobody knows what she's getting into when she starts to hatch: her parents might have situated her in a hackberry, a cactus, the bracken, or on a preposterous verge. For some birds, freedom is fun. Hackberry birds probably relish the irreversible freedom that comes with hatching. But to the gosling charged with getting herself off a zenith, freedom feels like conscription.

Someone a little older might protest—*What am I, loose change? Am I your provisional baby?* But three-day-olds are too young to be huffy and too young to know that anybody could be provisional. They are not, however, too young to be famished, or to know an affiliation so intense that the privation of that affiliation can send them hurtling off a cliff. When Mama was here, the nest was comfort and joy—now that she is gone, the nest is exile, like the Earth if the Sun flew away.

Some birds can have life for a song. A blue tit's mother feeds him a thousand times a day, and a secretary's mother brings her headless snakes. A cassowary's mother never flies away, because a cassowary's mother never flies. Some birds hatch into expandable nests of two thousand feathers or homey little saucer-nests, and their parents are fussbudgets and wouldn't dream of asking their babies to fly before they can fly. Their very first flight might be off a log—flying is like falling off a log! Little birdies flitting up and down from logs inspire elves and fairy sprites.

Asteroid children have it easiest of all, like the children in the Baptistina family, everybody crashing around the solar system together, all the heedless rocky little uncles and daughters and ancestors bouncing off the moons, shocking the worlds, leaving their marks, their Croccos and Drebbels and Blaggs. The Baptistinas may or may not have annihilated the dinosaurs, but if they did, they never gave it a second thought. They never lose their cool: when their family breaks apart, it just makes more of them. Where there was one cousin Alonzo there are now seventeen cousins Alonzo.

But geese never had any cool to lose, only warmth. Geese are more melancholy than rocks, being indivisible. Rocks do not pine; no rock experiences privation—or so we assume, only having seen rocks do possible things. Privation provokes impossible things: it provokes a song from someone who cannot sing, metaphysics from the materialist mind, and it makes little pedestrians to fly. Privation is as effective as someone running at you, clanging saucepans. Maybe at first when the mama flies away it seems like a test of forbearance, and the babies vow to wait till she returns. As her absence intensifies, maybe they make a more desperate vow—I will wait till you don't come back—till kingdom doesn't come.

However, if you get too hungry you will no longer be able to reach the food. The food is not going to pursue you; before it's too late you must bestir yourself. Famishment becomes acute, patience a curse; the heart goes white; the goslings never do not jump. The prospective audience, the lemmings and buntings, never stop to applaud; but it's hard to appreciate something barely appreciable—gray fuzzballs tumbling down a gray cliff—or something that so often comes to grief. Only the parents attend, waiting on the scree for the rendezvous.

After an absurd perpendicular may lie a broad plateau. The plateau itself might feel absurd at first—absurdly broad, absurdly easy. How absurd to be able to nibble bog cotton all day! But after enough time even bog cotton can seem ordinary. Even strawberries can become a given, even safety and surfeit and spelling, unless of course you try spelling in Yitha Yitha or Yorta Yorta. But if you stick to the givens and spell Englishly, you can get on with developing a forte, some consequence, some stature, and some priorities. Some people wish to gain entrance into the Clevery, some into the Forevery, and some wish to spend two weeks per year sweating healthfully in the Hummums.

Stature and perspective are supposedly commensurate: the taller you get, the farther you can see. But those who fancy themselves more omniscient than hatchlings may overlook the universe. Marooned on a verge in Greenland, tiny trickless geese look out at a moon pale like the ghost of a moon, at fog fading like interest, at dark clouds in front of darker clouds, at ice mountains and green valleys below. One thing they cannot see in this panorama is themselves—though this does not necessarily diminish the view. Yet if certain elements combine, they might get a gigantic inkling of themselves. On top of a mountain, looking away from the sun, you can sometimes see a massive shadowy shape of yourself, surrounded by a halo. A hatchling, a Viking hefting a dragon ship over his head, an ermine in her white winter coat are equally magnifiable—no one is too little to be magnified, no one too big.

In stories, if somebody does something deathly dangerous, like stepping off a cliff, then a bridge materializes underneath his feet, or he is changed into a sea eagle. Courage is validated by a presto-chango and the presto-chango is always convenient—the hero is never transformed into a hippopotamus, for example. But with the barnacle goslings, it is a no-chango situation: once the nestlings step off the edge they just become falling nestlings, subject to whiffles and outcrops and disappointment. The air suffers fools gladly but the ground does not.

Wishers they were birds are not thinking of nestlings, their winglets barely dried, obliged to take a mortifying fall—not gossamer babies hatched on top of the world, having to jump off, having to be so brave in their nonage, having accrued almost no time or material, having within their purview merely the moon and the sun, the mountains and valleys and, seemingly without a source, the wind.

The willy-nilly wind: if there is validation in nothing else, there is validation in the wind, in the weather, whether one experiences three days or three decades of it. Once you have felt a breeze fluff your feathers, a drizzle on your downy head, you *have weathered.* Like "wrestle" and "whistle" and "fly," "weather" is an action verb. You are not just a passive recipient of weather, not just the pelted, the drizzled, the fluffed.

This includes the most provisional among us, as well as the least: we Weather. We Winter, we Summer—but this is not true of the bird who never hatches, who never emerges from her chamber of ease—we Spring and we Fall.

ESPECIALLY THE ZEBRAS

Two giant Aldabra tortoises with wide dark eyes are on the move around their scrubby domain. The front one stops. Not willing to participate in gridlock, not wishing to change lanes, the rear tortoise climbs over the front one, like a Fiat driving over a stopped Fiat. The one underneath objects wrathfully and shakes the surmounter off. Another giant tortoise hauls himself over on thick pigeon-toed legs, and the three of them face each other like the complacent lobes of a shamrock. A fourth tortoise in the shed has lost all his go.

Two female giraffes arrived in Tulsa this week, one from Missouri and one from Kansas. For now, a fence separates them from the male giraffe, but the fence is low enough so they can, as the sign says, "interact with his head." They sniff his head, nuzzle his head, take cues from his head. At first the two of them were startled by all the roaring (the Tulsa Zoo is next to the Tulsa Airport), but then they looked at Samburu's head, which was as cool as a cucumber. He was born here in 1992 and knows that jet planes are as natural as wagons full of grape-eating children and the intermittent ditty of the carousel. A tall platform is being constructed where the carrot-bearing public will be able to climb up and interact with the giraffes' heads. It must be a thankless job to be the body of a zoo giraffe, always transporting the far more popular head around. But someday soon, Lexi and Pili will be let into the main yard, to join Samburu in the circulating shadow of his canvas umbrella, the three of them forming a three-giraffe sundial.

It is so hot the leaves flash like mirrors. The cicadas get riled up, and cottonwood fluffs around. Two little gray Sicilian donkeys marked like hot cross buns are in a stupor. A spindly little girl in the Frogger Jumping Harness dangles in the air like a pixie, too light to actually touch down on the green bouncy mattress. Gravity can be a good friend, but I have noticed that he plays favorites. Three chunky zebras are safely grazing; the cheetah lives behind an unleapable fence. Oklahoma looks like Botswana, especially the clouds and the zebras, but you can tell you are in Oklahoma because the cheetah is visible.

In the Children's Zoo, a Red Wattle hog mixes himself some mud with his snout—start with some dirt, add some sump water, add a little more dirt, mix mix mix—and plunks himself down in the oozy sludge, his big notched ears flapping forward over his eyes. His explanatory plaque says that Red Wattle hog numbers dwindled when people found fattier hogs for making soap. It also says that Red Wattle hogs have some wild relatives in Texas—the wild Texas relatives must have quit the soap-making business to pursue mud-making full-time. In the Contact Yard I try to make contact with an extra-wide goat, stroking her rusty-wire hair. She does not appear to notice. One little crackpot goat is running around butting other goats in the rear, and they definitely notice. Here is someone who knows how to make contact that sinks in.

A grizzly bear has her nose pressed against the window, watching the children watching her. Apparently she was found at the same dumpster or campground three times, and the rule is three strikes and you're out, and they know who you are because they tattoo your lips. Grizzly bears have such a powerful sense of smell that, even if tranquilized, crated, trucked three hundred miles away, lectured to stay put and eat the local lady-bugs and huckleberries and wild potatoes, they might show back up behind Buckshot Bessie's Bar and Grill, ransacking that peerless dumpster, that consummate dumpster, that dumpster surpassing all dumpsters. The polar bear died so the Tulsa Zoo had room for a recidivist.

A Scottish Highland cow with shaggy red bangs and tufty eyelashes is breathing so heavily her long curved horns rock back and forth. The word "animal" comes from the Latin word for "having breath"—*animalis.* Every time I walk by the alligator snapping turtle's aquarium, he is slanted up his slope of rocks. I don't understand why he's always in the same position till I see him stretch his head up to the surface to breathe. His stony-plated armor looks fortified for fending off Gog and Magog but here, submerged in his tank in Tulsa, all he needs to fend off is a small fish friend and death, so he positions himself for convenient access to air.

Close to a mini-bayou graced with two alligators and a cabana, there is a triptych of small embassies: SWAMP, BOG, and MARSH. The barking tree frogs representing MARSH are stuck to the wall. The tiny BOG ambassador is lying exposed in a dish of mud; the description of bog turtles says "They are very secretive." She is not alone among secrets these days, having been transferred from a ferny bog into the public eye. Appearing for SWAMP is a spotted turtle that could fit into the palm of my hand, swimming hard into the window, making the water churn around his fake mangrove roots. As I walk back and forth, he follows me, climbs onto his minute bank, falls pinwheeling into the water, paddles back to the glass, cranes his neck toward me. If he were an ambassador from outer space instead of just a swamp, I guess we'd all be amazed; or if he were a man-made turtle programmed to paddle mechanically after a person.

In the back of a heptagonal snow-and-sky diorama, furnished with dead-leafy branches, a snowy owl sits swiveling his head to the left, around to the right, around to the left. Two different people who pass by inform their companions that owls can turn their heads all the way around, but they wonder why he doesn't do it. A hefty girl pounds on the glass and shouts, "That is a sexy owl!" Two teenagers in cutoffs conduct an antiphonal conversation involving someone else on the other end of a cell phone: "Have you been in the truck?" "I've never been in the truck." "He says he's never been in the truck." A caravansary of preschoolers is ushered through. At first they are excited that the owl has his very own bath, but then the excitement turns Aristotelian as they debate whether it's a bath or a shower. Preschool does not mean pre-intellectual. The sign says, "The snowy owl's range includes the entire Arctic Circle."

A few pre-intellectuals are strollered past the dwarf antelope huddling in the shadow of her twelve-foot wooden fence. They seem to be the only people not asking, "What's that?" "What the fuck?" "What the hell is that?"—the only ones whose interest in the animal is not deflected by the answer "a dik-dik." One little boy waves at the antelope silently; two dark-haired baby twins in yellow dresses stare, open-mouthed; a small girl in gold-heart sunglasses cries out "Goodbye" as she is wheeled away. This all helps me understand why only two-year-olds are allowed to go on a Camel Tour:

1. Riders must be 2 years old and able to hold on securely
2. No kicking, biting, spitting, or other offensive acts toward the camels
3. The load limit will not exceed 300 pounds

Two camels lie under a tent, patiently wearing blue seats. Who is worthy to be carried by such faithful creatures, except nonviolent two-year-olds under three hundred pounds? Babies should qualify too, because of the hallelujah quality of their minds, but they might fall off.

Outside the chimpanzee house, there's a red metal outline of a chimpanzee bolted to the concrete. It looks like a chimpanzee cookie-cutter; a woman has her daughter get inside it, to compare physiques: "Scooch forward, stick your butt out." A chimpanzee has very long arms and very short legs, so to fit in the outline a person has to contort herself. But none of the actual chimpanzees I see would fit in the cookie-cutter, either, not Bernsen the little pirouetting one, or the huge one, Morris, who catches flying bananas and lemons with one hand.

I seem to be living in the age of anatomical expertise. Signs at the zoo say things like "Langurs have tails up to 43 inches long," "A chimp's muscle tissue is denser than ours," and "Elephant tusks are incisor teeth." At the Elephant Demonstration, visitors ask, "How many bones do they have?" "How many toes do they have?" "How many teeth do they have?" An elephant has four molars, like a dead elephant, but a dead elephant can't hold a candle to Gunda here, enjoying her broccoli and her hose bath, her particular freckles emerging from the mud.

Anatomy is an animal's identity—except, when I

talk to the people who know the animals well—the elephant keeper, the giant-tortoise keeper, the grizzly-bear keeper, the chimpanzee keeper, the sloth keeper—they don't talk about bones or teeth or toes or tails. All they talk about is personality—how distinct the elephants are from each other, how the tortoises differ in their fearlessness, pushiness, irascibility, love for togetherness, mothering styles, interest in people or antipathy to them. Three Fiats with homogeneous bodies will yet behave in heterogeneous ways, one creeping hesitantly to the pharmacy, another zooming to a toga party, and one, having forgotten to fill up, slowing down and stopping.

You might think a sloth would have little behavior to differentiate it from its fellows, but Ty for one likes to sleep in a bucket, and, after months of sweet-potato blandishments, Ty does not mind the keeper lifting his long curved nails up out of his bucket into the air. The only animals interchangeable like dollars are the ones not indigenous to this planet—the GloFish®. So that you must purchase these fluorescent fish from a retailer, the manufacturer makes them sterile. Perhaps something in this process has warped them: all of the GloFish® at the Tulsa Zoo have the same personality—deceased.

Once, the elephant keeper worked at a zoo where they'd take the elephants for walkabout in forests and meadows, and he said if you didn't keep an eye on them you'd lose track of them, their footfalls are so silent. It's true—I watch the male elephant trudging around his dusty yard, twirling the end of his trunk, swinging his tail back and forth, fanning his triangular ears, but I cannot hear him, gravity's darling, any more than I can hear the dark-purple butterfly fluttering back and forth in front of me. One got all the mass, the other got all the autonomy, but they are equally inaccessible to the ear; there is more than one way to escape scrutiny.

Before I leave the zoo, I pass a sign that asks me, "What animals do you want to attract?" I consider, and decide that, of all the animals I have seen here, I would most like to attract the Red Wattle hog and the extra-wide goat and the lithe green lizard I saw resting his chin on a log. I enter the Habitat Garden to learn how. However, I find the Habitat Garden to be strongly skewed toward the attracting of toads. "Overturn a clay pot in a damp area for toads." "Large rocks and rock piles make good shady places for toads." "Attract toads with a slow-dripping faucet." Although I am glad to learn how to sweeten the life of a toad, I leave the zoo not knowing how to be someone the hog, the goat, the lizard, the dik-dik, the gibbon, the saki monkey, or the snowy owl want to be close to. As I approach my Fiat in the parking lot, two little brown birds who had been sheltering from the sun underneath my car, two free little Oklahoma birds, fly away.

PRESTO

Early on, there was no word for "groundhog." Neither were there groundhogs, or grandmothers, or event coordinators. There were events but they were uncoordinated like the Tunguska Event. There was nothing, but no word for it. In some ways it must have been nice, all that wordlessness, because sometimes now you meet somebody and all you can think is, *Please stop talking*. Our planet has become so much wordier than the other planets, although there are respites if you hang out with eagles or angels. Eagles never explain anything, and angels are no more voluble than they are visible: visibility is not their shtick.

But once in a while you meet a visible voluble being and you think, *Please do not stop talking*. This would never happen on Neptune. Also, with all the flowers and words for flowers, we can indulge in wildflower gossip, about how the golden alexanders are bolting, how the tansies are being insubordinate again, how the mad apples are creeping up on the obedient plant, how the purple archangel was boogying in the wind last night. (The purple archangel is anomalously visible.)

Wildflower scandals are talkable things. There are talkable things and untalkable things, just like there are beable things and unbeable things, though not everybody knows this. First-graders want to be beable things—astronauts and firefighters—when they grow up, but children in nursery school want to be owls and cars and sandwiches. They are expatriates from somewhere far-out, and they are still trying to talk about untalkable things, like: How fast is time? What is time plus time? How many jiffies in an hour, how many soons in an afternoon? How many fasts does it take to make a slow?

Once, I found a book of piano music by Brahms and played an intermezzo slowly, very slowly, lentissimo. I thought it sublime at that tempo, but later I heard someone on the radio playing the same intermezzo fast, fastissimo—she could have played it four times while I played it once. In this case it took four fasts to make a slow. When my father was nine, his parents had him sing "The Holy City" for church, accompanied by his four-year-old brother, Danny, pumping the pedals of a player piano. At first the song rolled right along, but then, as Danny got tired, he pumped the pedals more slowly, and Benjie had to sing more and more slowly until, right in the middle of a word—"Jerrr-ru-u-u-sa-le-e-mmm, Jerrr-ru-u-u-u-u-u"—the song stopped, and the congregation roared with laughter. Then Danny got a burst of energy again, and the song resumed and slowed down again and stopped again.

Time can seem just that erratic, with prestissimo, lentissimo, and us always having to scramble to keep up or slow down to a mortifying pace. But slow or fast or spasmodic, a song always ends. An hour is like a song in that it ends. For a while there were a lot of Triassic types going around. Montesquieu made a name for himself, as did mosquitoes. The tribes of the Warm Springs moved from their winter village to their summer village and back. Passively the icicles grew, passively they melted. Some people lost their good reputations, some people lost their canoes. Some cowboys lost their cows and then they were just boys. Nations formed, nations raged; meanwhile the giraffes were so well behaved we forgot all about them. Some people, you could tell, spent their evenings reading pamphlets entitled *Why Won't They Listen?* Punctiliously the clocks kept ticking. There was a rolling hello as people appeared, and a rolling goodbye as they disappeared. The generality was perpetually replenished, but you don't really say hello or goodbye to a generality.

For all the things that did happen, many more things did not happen. Many people did not appear; many people appeared but did not flower; Meister Eckhart flowered but did not drive a Maserati. The event that flattened all those trees in Tunguska did not flatten the folks in Delhi or Quito or Winsted, Connecticut. Even after Maseratis appeared, they did not impinge on some people's consciousness, just as nuthatches do not impinge on some people's consciousness. Some people never got around to reading *Moby-Dick*, so they never realized that everything in their lives had been preparing them to read *Moby-Dick*.

And all this happening and non-happening has brought us to the present hour, when you have cats running around younger than churches and trees standing there older than churches. And we are still innocent of time. After all this time, to be innocent of time. Though maybe there are some things you can't realize until your name is Grandmother, like the tempo of time, like those grandparent types who go around insisting, "It goes so fast." Maybe all those slows add up to a fast. Maybe it's like we're sitting in the garden with our little cups of water and some excited person keeps rushing around telling us, "Water is blue, water is blue, can't you see, water is blue." But we peer into our cups and the water doesn't look blue—it looks clear, like a good little diluter. We add it to our whiskey, we add it to our milk, and they get clearer, weaker, waterier. There is just one thing that water does not dilute, and that is water. Water plus water plus water is blue. Sinking down, you see how very blue.

O LATITUDO

I loved being young,” says the chorus in *Heracles*, written around 420 B.C., and to this day people are still chorusing about how they loved being young, insisting that someday young people, too, will have loved being young. Nobody goes up to old people and says, *I loved being old*, or if they do you can't hear them or see them. I know of a bear who did not love everything about being young, the bear who, his first winter, built a den for himself, but it wasn't big enough, so all winter long he hibernated with his bottom sticking out.

Platypuses could say to us, *I loved being young*, since compared with platypuses humans are whippersnappers. Platypuses may condescend to us but we must not condescend to them. If we want to condescend to somebody we can condescend to Petunia. Petunia is a ridiculously young flower, having been cultivated in the nineteenth century. (Petunia's parents are wild old Argentinians.) But though she is young and inexperienced, I have seen Petunia rallying after a hailstorm, putting herself back together. I imagine if the supervolcano a little south of here erupted, there would be the insouciant Petunia afterward, shaking the ash off her purple.

Breezing around over a supervolcano, the dragonflies of Yellowstone are so insouciant you might think they were superyoung. However, they are three hundred million years old, and over that time have seen plenty of supervolcanos erupt. Maybe that is why their heads are all eyeballs. Anyway, rather than organizing a superbucket brigade in preparation for another supereruption, the flame skimmers, cherry-faced meadowhawks, and mountain emeralds are zooming backward, zooming forward, zooming up and down, zooming in place. They possess the insouciance of the old.

The supervolcano has a supersecret underneath the surface, seething magma and hot mushy crystals. On the surface, that secret is expressed in the bluest pools, the most experimental rocks, the burpiest mud, and the rainbowiest steam. Looking toward the Grand Prismatic Spring from far off, you can see its prismatic steams rising into the air, red steam, green steam, tangerine steam. The steams are the supervolcano letting off steam, and the colors are the colors of the swimmers in the spring.

Of course, in the swimming-pool biz, the tradition is to keep the water "safe," with a pH of 7.4 and a temperature between 83 and 86 degrees Fahrenheit to make it "comfortable for your swimmers." However, it depends on what type of swimmers you want—if you want fragile, big, loud, pink and brown swimmers, then that's a good range. If on the other hand you keep your pool at 450 degrees Fahrenheit with a pH of 2, like lemon juice, you will get flinty little green and orange swimmers who never shout.

So extreme pools attract extreme patrons; so much for toning things down. Still, the supervolcano is probably at its most brilliant when not erupting. A kept secret is an engine of invention, and Yellowstone's supervolcano is flamboyantly secretive. Because it does not eject its secret, its secret is ever imminent, effecting pools so turquoise, so russet, so lime, waters so swashbuckling, and rocks so imaginative—those beehive-looking rocks and mammoth-looking rocks and urchins and elephants and hoodoos. Landscapes with no secret can be a snore.

Hoodoos are imaginative rocks; they are people-looking columns, each with a hard little limestone cap that keeps its softer, sandier torso and legs from eroding. Actually the hoodoos of Yellowstone are not hoodoos but boulders that tumbled down a hill. However, they resemble hoodoos and hoodoos resemble hooligans. Hoodoos resemble hooligans both externally and in a deeper way: as a hooligan is not just a hooligan, so a hoodoo is not just a hoodoo. That is why hoodoos and hooligans are never disappointing.

Sandhill cranes fly up from Arizona to Yellowstone to have their babies on the supervolcano. Their babies are destination babies. The cranes must know what they are doing because, like the dragonflies, they are very old, dating from the Pleistocene, when some extinct humans appeared. As it turns out, those humans were fifth wheels—who misses them now? Who cares if a fifth wheel falls off? Anyway, maybe they went extinct because they were impatient. The patience of the sandhill crane is exemplified by that mother sitting on her egg in the spring blizzard. While the snow piles up to her wings, up to her neck, up to her eyes, her baby bird never even knows it's snowing. The mother crane is patient, like Monteverdi, who had terrible headaches but you'd never know it from his music. Here's to all you mothers out there sitting on your babies in the snow.

Our mother the world is very old, and like a lot of old and wild parents—Abraham Lincoln and God and the parents of Petunia—she is quite permissive. She gives us plenty of latitude and longitude and endless examples of what is permissible. The bubbly fresh springs give us permission to think bubbly fresh thoughts; the sulfurous pools give us permission to think stinky sulfurous thoughts. Winter gives us permission to think dark icy thoughts, especially if our dens are not big enough to accommodate our bottoms.

The slug gives us permission to be sluggish, the bump on the log to be a bump on a log, and the bison gives us permission to go around with sprigs of juniper tangled in our hair. Authorities like coyotes give us permission to possess some bite, and authorities like dragonflies entitle us to zoom in place in the presence of great explosivity. Petunia authorizes us to be ridiculously young and the platypus authorizes us to be ridiculously old as well as ridiculously ridiculous.

The erupting volcano authorizes you to erupt, even if you are no longer three, even if you are a million years old. Exploding is legitimate and can be fun at any age. However, after years of watching Yellowstone *not* explode, I wonder if containing one's supervolcanic forces, so they bubble up to the surface in mud or watercolors or arpeggios or spumoni or whatever one's medium is, might be even more fun than exploding. Maybe even better than ejecting a fiery superplume of lava into the sky and blanketing Montana in ashes three feet deep—better than burning down Wyoming, with all its beloved forest-floor animals—is containing the volcano, and consequently composing a gassy, burpy, muddy Ode to Joy.

THE GIFT

[Tigers] can live alongside people, invisible and unnoticed.

—Fiona Sunquist and Mel Sunquist, ***The Wild Cat Book***

Along with the binkies and blankies, the swaddling cloths and sleeping sacques, the tiny bath towels with owl-face hoods, the melamine mermaid plates, the cloth books about ladybugs, the pods for sitting the baby in, and the mechanical swings that play lullabies and whooshing sounds, there is always another gift for the baby. Mostly she is showered with fun gifts, safe gifts—safety nail-clippers with built-in flashlights, safety cribs with slats through which her head cannot fit, safety duckies whose bottoms read "HOT" if the bathwater is too hot. The world being full of hot water and three-foot drops, the baby draws out all the fatherly-motherly-uncly-auntiness in people, who come from towns away to bootie and bonnet and duckie her.

But somebody always brings this other gift—this extra, anomalous gift—this gift that is not third-party tested and is in no way compliant with federal safety standards. Somebody gives the baby a tiger. Not a toy, but a tiger, immense and silent and, because of its stripes, invisible. Those dark stripes break up the tiger's outline, so it just looks like some fragments of sunshine over by the sofa. It is easy to ignore the invisible gift behind all the other gifts, but every baby gets one: the Panamanian babies, the babies of Ypsilanti, and the babies of the past, the ones who wanted for nothing and the ones who wanted for everything, including time. *I had just arrived in the fourteenth century, and it was pulled out from under me like a rug.* Such tiny flameouts had their tigers too.

Parents can forget that their infants have something that was extra-parentally provided, and the lullabies make no mention of it—"Hushabye, don't you cry, go to sleep, you little baby [with a tiger]." But to every baby delivered into this world is delivered this wild animal. Now, normally when we call somebody wild we are thinking loud and crazy—the person dancing on the table rather than the person in the corner who would like to go home and stew vegetables. But with animals, wild usually means reticent. Throw a party for lynxes and herons and nutrias and most of the guests will be hugging the walls, barely flicking their tails, hacking up the odd hairball. You might see some otters running around with trash cans on their heads, but most wild animals are not party animals. Most wild animals are wallflowers.

It usually works out pretty well, the worldwide balance between wallflowers and attention-grabbers, because we have only so much attention for the grabbing. Imagine if sand dollars were trying to get famous, too. The only trouble is, sometimes it is hard to remember there is anything extra-cultural out there, like stars. Maybe you went to university thinking you'd learn about the universe, only to find the stars swept under the rug and culture dancing on the table. When it comes to the nations, the stars are really taboo; if you're ever out hobnobbing with the nations, don't mention the stars—the nations are very sensitive and do not like to think about anything that is not nationalizable. Talk about flags, statues, polls, wars, money, summits, leaders; talk about tsars, not stars. Save the stars for the elvers. Elvers are no provincialists. You can talk to them about anything.

Like parents, the nations can forget that their subjects have anything they did not themselves provide. But the tigers dispensed to babies are no more national than the moon. Even the Spanishest baby has her universal tiger—and not just briefly, while she is a baby. Spanish people, Syrian people, horn-players, tortilla magnates, persons named Albertine; needleworkers, noodleworkers, big George carrying little George down the hall—all are tailed by those common-rare creatures.

The tigers are common because everybody gets one and rare because everybody gets only one. One can be such a disconcerting number. For people who aspire to be anchovies—members of a swarm intelligence—having that single, solitary, unshareable tiger is unsettling. Also unsettling is not being able to look into the tigers' production and distribution. It's not like you see trucks backing up to tiger farms, loading up, then barreling down the highway and offloading them at stores where they get shelved beside the car seats and formula and diapers, where you can choose the one whose disposition and provenance suit you best and then coax it into your cart and up onto the conveyer belt at checkout. You can't choose your tiger or your baby's tiger and you can't trade it for a more chicken-sized animal.

It can be hard to stop thinking like a customer. However, a tiger you did not purchase is following you. She is intractably free—she costs nothing and takes no directives—and she leaves you intractably free—she is not a nanny animal. So you want to be a rumba-dancer: your tiger will lie nearby, impersonating sunshine, while you practice the box step. Paddle through the Great Dismal Swamp and she will slog behind you, impersonating ripples. Move to the suburbs and she will become a tiger slumming it in the suburbs, stepping gingerly around the tricycles on her big soft soundless paws. It can almost make you think there are no wrong roads.

Billboards along the highway that read "LIFE IS A MIRACLE" always feature a starry-eyed baby rather than a weary middle-aged person or a cow or a prickly pear. Of course, the cows are already employed on other billboards, but prickly pears are surprisingly underutilized. That middle-aged people never get recruited for those billboards is not surprising. We are no longer very starry-eyed. We are earthy-eyed. Maybe we could be on billboards that said "LIFE IS ENDURANCE" or "IT'S A MIRACLE I'M STILL ALIVE."

We who are in the middle of the miracle, who have

found a place in society, have mostly forgotten about our extra-cultural, extra-political, extra-commercial tigers. Not that the tigers were any help, being so good at being forgotten. Beavers make their noise with their smooth, broad tails, and Tom Waits makes his noise with his bull-horns and tin cans, and tigers make their silence with their big pillow-padded paws. In snow they keep to the deer paths, so their paws won't go crunching through the crust. Occasionally, after a strange dream or a Tom Waits song, we get an intimation of them and try to turn around and look straight into their fiery amber eyes. But usually we just keep turning and turning.

Still, there's that one extra-green streak of grass running through the yard like an innuendo. And once in a while there's a quiet that feels not quiet like a mouse but quiet like a tiger. It makes us feel that something is afoot. Usually, people consult up: little people consult big people, big people consult professional people. Some people consult Jagannath, the Lord of the World, throwing themselves under the wheels of his cart. Rarely does anyone consult the babies, throwing himself under the wheels of their strollers. But babies before they become personable—with their trances, their abandon, their seriousness, the leeway they give us, letting us dress them in frills or duckling suits or banana costumes; with the way they fix their eyes on something we cannot see—maybe the babies are regarding that one anomalous gift behind all the other gifts. In this way the babies are better than professional, better than lord of the world, having so lately been ambushed by life.

GREEN MAN

For all the trumpets and slippers and stars popping out of the ground right now, you'd think the Earth was stocked with flowers, that underneath the surface was a flower warehouse, pallets of poppies stacked on pallets of phlox; and if you could only find the entrance, you'd discover employees down there, fluffing and freshening the inventory, forklifting the pallets around, and slipping the flowers up through the dirt as soon as it gets Aprilly warm. Where are things housed if not in warehouses?

But the Earth is not full of flowers, just schist and samarskite and skulls, and hardpan so infernally dense it cracks your pickax and then your crowbar, shatters your jackhammer, your million-pound excavator, and finally your two-million-pound excavator, after it tries to pull the smaller one out, spurts oil from its splitter box, and kicks the bucket too.

Earth is a frustrating substrate to live on, by Gog—not just for jackhammers and excavators but also for little mesquite beans plopped down in the desert. Whose fiendish joke is it, to drop them onto a griddle and bid them grow? However, it is not a fiend but a cow, and the cow is not making a joke, just a turd, then off she ambles with all her cares. The bean must be its own bidder, its own shovel and shoveler, and its own yardstick, measuring the distance to the water with itself.

Now, if distance is your only problem, you can just plod your way to success, like a hexapod. But often there is also the problem of direction—the things you need being scattered willy-nilly through the world, at fathomless slants, at terrible removes. Finding them calls for not only determination but also derangement. Mesquites send their roots hazarding in every direction, turning the earth upside down trying to find a drink. (The mesquite tree has been called "the devil with roots," but this is obviously a misnomer, since devils can only travel in straight lines. Japanese zigzag bridges are built with this in mind.)

If only the roots could sense where the water is, like witching sticks; but roots have no more premonition than yardsticks. Into the ground they grow, whose only power is stab, displacing the earth particle by particle, tedious and slow-straining and unmagic: if Earth be thy substrate, then Earth be thy substrate.

After you've been stabbing in the dark for a while and you stub your roots on hardpan and it seems as though you haven't made the slightest progress—are everywhere blighted, everywhere foiled—it might be tempting just to conk out. *Enough is enough, by Jolly! Why must I spend my life blundering around in the dark, looking for something that might not even exist?* Answering voices will eagerly confirm that water is a far-fetched notion, that you can relinquish your tormenting strain; they will happily receive you into their confederacy. *We once felt that juvenile strain toward water, too, before we wised up and learned to appreciate the available. Give up your anguished unrest, your* desiderio, *your solitary wrestling with rocks.* These are the dogmatically dead, who, because they have not dug down very deep, think they are very intelligent. They pass the jug of dust, smacking their mandibles.

But after enough time, hope is no longer so diffident and deferential, so easily suppressed if you come across a tempting alternative to life. When hope is an infant, you can swing it along in a bunting bag or leave it swaddled in the banjo box. But with time and exertion, hope waxes huge—huge as a Pictish king. Even if you tire of his outlandish campaigns, ain't no Pictish king going to stay swaddled in the banjo box.

Of course, some situations call for more rebellious hopes than others. Living out in the voluptuous Atchafalaya, tupelos have no need of hope, with showers falling into their laps every other hour. The Earth is their tuffet. Senita of Sonora, she can just sleep until it rains, then grow her little rain roots to sip it up. And air plants like Spanish moss needn't make any contact with the ground whatsoever; they are loose as leprechauns; they drape themselves over tree branches, dangling their roots, drinking in the fantabulous fog.

Some people, when they open their doors in the morning, have scores of angels surging in, so many that they have to contrive minor favors for them to perform, such as hamster-petting and thimble-polishing. Other people, if they have an angel at all, he's an angel on the run, an estranged angel, and they have to hunt him down in the wilderness, hiding behind boulders all night, to try to ambush him and wrestle him to the ground, stomping down heavy boots on his colossal pearly-feathered wings, demanding, for God's sake, a little help.

Such an adversarial angel is the Earth, for some

of its residents, its help so unavailable. There is fresh water under the ocean, but whose roots could reach it? The Ogallala Aquifer underlies Texas and Kansas and Nebraska, but unevenly, in places a hundred feet below the surface; in others four hundred feet down. They say difficulty is an invitation; does that mean the more the difficulty, the more the invitation? Is impossibility the deepest invitation of all? Mesquites have been found with their roots grown right through hardpan, grappling the planet: *I will not let you go until you bless me!*

Because of his utterness, the mesquite is considered a chop-down tree and rarely gets invited to lawn parties. Ornamental trees know how to get invited to lawn parties, how to cultivate influential friends: they do this by making their friends feel influential. They send a branch in a gauche direction or grow eleven extra leaves on one side, and call, *Yoo-hoo, attend, my contours are in danger!* and friends come running to remove the offending material. The dandy gets waited on, twig and trunk, and the influencers get to tend to a plant with fixable problems, problems of punctilio, problems of contour, problems of form.

The mesquite makes nobody feel influential, because it is impossible to influence him. You can't stop his everlasting littering or get him to stop scratching your cattle with his three-inch thorns; you can hardly buy or sell him or plant him or transplant him, or kill him: administer your bulldozers, heavy chains, fire, he will laugh them off, for he is importunity incarnate. Chop down the incarnation, the incarnation returns. Hack off his visible self, dance on his grave, and soon you'll be dancing in his sprouting green hair, then up in his strong arms, and how embarrassing, unless you are a turkey, to be seen dancing in the arms of a tree.

This is why only turkeys and other gallybaggers enjoy mesquites. The lacy shade and bittersweet beans of the mesquite attract wild pigs and pocket mice and porcupines and a few dotty Texans and screech owls. For screech owls, a good party is not one where they can see and be seen but one where they can see and not be seen. In the feather-resembling bark of the mesquite, little screech owls can impersonate nobody and ponder the riffraff.

This is how the mesquite ends up with a confederacy after all: not a confederacy of cadavers, or one of influencers, but a confederacy of enjoyers; not a confederacy it joins before it has a self but one it elicits with a self, a flinty self with vast underground contours. Errancy is onerous and errancy is freedom and errancy is exercise. If you have an inerrant javelin that never misses the mark, you need to throw it only once; but with an errant javelin, you grow strong with all the overthrowing, underthrowing, and rummaging through the junipers. After all his fumblesome toil, you might think the mesquite would be a grim man; but if he toils into water he's a green man, merry-mantled, foresting the desert. Like all beans, the mesquite returns nitrogen to the ground, transforming a scorchy waste into a greeny glade for gaga mice.

But the weather never goes away. Aquifers can fall five feet in a year, and sometimes all you get from a storm is wind. Drought forces the mesquite to throw off his leaves, his bonhomie, and go back to rummaging; and he may come across water, or he may send his taproot down to tarnation and still be sublimely wide of the mark, errant to no end. *To fathom and fathom and nothing to find*. The freedom to not find water might sound miserable compared with the freedom to fly babblerlike to Ouagadougou or run houndlike on the hummocks or go vanquishing the Visigoths. But the freedom to dig his own disputable way—that's the freedom the thorny little tree was given. And if 160 feet of earth intervene between him and nothing, and he discovers down there not the Ogallala but an abominably yawning copper mine—still he has the freedom to house in his heart that dissident green, until the last stab—to never mistake dust, though he swallow barrelfuls of the stuff, for water.

THE MODERN MOOSE

Recently, some modern animals have been reconsidering their attachment to the Earth. She was a lot of fun in her honeyed youth, but she's getting sick and seedy now, temperamental and pockmarked and tired. They find her decline kind of depressing, kind of repulsive. Before she is totally moribund, they are looking around for other options, redirecting their attention to places like Mars and the Large Magellanic Cloud, which contain untapped riches of rare materials and are unaffected by banana blight. In fact, such moderns have a hoof or two in heaven already: if they say yes to the Earth, it's an equivocal yes, easy to disavow as soon as they can blast off to a better where.

But as ravishing as Mars may be, not all modern

animals seem so eager to leave the Earth. Some are still entangled with seas and trees and Russian tundra; others seem entirely indifferent to the idea of space settlement. Take the modern moose, for example. Has he sent scouts to the moon? Has he shown any interest in starships? Has he ever practiced grooming himself in an antigravity gyroscopic device? No, no, and egads, no. Nevertheless, Mr. Moose is as modern as Mr. Musk and must be equally respected. In his book *Modern History, or the Present State of All Nations*, published in 1744, Thomas Salmon noted the concerns of the moose: the moose likes to chew on young shrubs, "but mostly, and with greatest delight, on water-plants, especially a sort of wild Colts-foot and Lilly that abound in our ponds, and by the sides of the rivers, and for which the Moose will wade far and deep."

With the magnitude of his antlers, it's not like the moose could ever be flighty anyway. The first time bone starts coming out of his brow, the young moose might think it will turn into something moderate and flattering, something like a pillbox hat. Maybe his horns will be trinket horns, party horns, flirty horns like the giraffe's, or sleek Armani antlers like the pronghorn's. But the bumps grow into spikes, and the spikes spread and branch and keep growing, past trinket, past flirty, past flattering, and far past moderate. (Moderate horns are for moderate species; moderate species get very excited about moderate horns.) Finally they grow past preposterous: on his head the moose carries branching antlers so absurdly heavy he mustn't lower his head down to the ground, for fear he'll never raise it up again.

Seventy pounds of antler seems like an affirmation somehow, an exaggerated weight attaching the moose implacably to the Earth, saying, *Yes without a question, yes with all my heart. Yes if it pitches me face-first into the mud, yes if the rest of me withers, yes if my yes gets splintered, or broken, or deformed: yes and yes and yes again.* When you have wings, your hope can be elsewhere—up in the sky. But wings of solid bone scorn the idea of flight: your hope must be here.

Of course, the moose didn't choose his yes any more than Respighi chose his. It is his hap to be born—to come out of the mama onto the Earth—his hap to be rejected by the mama once a littler moose comes out, his hap to bear a staggering affirmation on his brow. A real yes weighs you down, like a woe, like those lunatic vows of lovers, to be kept even if one finds a better who. *I,* Alces alces, *take you, Earth, to be my planet, to have and to hold, from this day forward, for better or for worse, for richer, for poorer, in sickness and in health, to love and to cherish; from this day forward until death do us part.* Yes to the pond where the water-plants thrive, yes to the pond where the water-plants fail, yes to the pock where the pond used to be. Yes to you healthy, yes to you sick, yes to you blooming, and yes to you stricken. Though you have seen better days, though you no longer delight me with Colts-foot and Lilies, yes to the Earth, my Earth, for I do not hope to find a better where.

SLEEPERS AWAKE

For all the talk about Mercury being mercurial, Venus is no less versatile. She's the evening star, the morning star, a bog plant, a burrowing mollusk, and the goddess of love. And before she became the goddess of love, she was the spirit of kitchen gardens. She who presently destabilizes people was once employed in stabilizing radicchios. The pace of life was plantly, Venus watching, waiting, glowering at the moles, succoring the Brussels sprouts, so there would always be something to feed the visiting Belgians. In this capacity, she was a goddess with no cachet—or all the cachet of Wee Willie Winkie, making sure the vegetables were in their beds. But then, in Venus's subsequent life, it was lucky she had been a gardener, because she knew how to console herself with lettuce when she lost her lover. It's good to have experience with lettuce, endives, chervils—something besides just love.

These days, we have to be the goddesses of our own gardens, dealing ourselves with the undermining moles and overspreading purslane. Being a goddess can be tedious. Venus, after all those centuries taking care of parsnips, is busy uprooting hearts, tossing them into the air, tying them to seesaws, loop the loops, Turbo Drops. Perhaps to keep herself flexible, Venus has retained some of her other offices—those of clam and carnivorous plant and planet. As a planet Venus is sluggish and hellish. Her day is longer than her year, and her clouds are composed like battery acid, keeping her temperature at around 870 degrees Fahrenheit, day and night. Her rain never reaches the ground, her clouds never gather because they never dissipate, and in this aspect—the weather aspect—Venus is the opposite of changeable, while the weather on Earth is some kind of poltergeist, with its tossings and fulminatings and feints.

Over the eons, there has been considerable conjecture about the otherworld. But since we have an otherworld just one orbit away, it seems like we could do a little less conjecturing and a little more scrutinizing. On Venus, little seeds of nothing are scattered in the ground, tended to by nobody, sprouting into numerous big nothings, laden with nothing. All the somethings on Earth, the umbels and peduncles and spurreys and spinneys and corks, are probably why everyone hangs out here, why the other planets are so godless. Venus on Venus would be as frustrated as Satan on Saturn.

Similarly sized and composed and situated, Earth and Venus are said to be twins, except one of the twins is very eccentric. Venus is like the sister who wants to be the Apotheosis of Everyone Else—empty like all the other planets. And Earth is the eccentric twin, eccentrically temperate, eccentrically inhabited, with her puggles and stingo drinkers and finches singing uncommissioned songs.

Except with time, Earth can come to seem as paltry as her sister, because repetition can divest things of their quiddity. Wake up to Trebizond every morning and it becomes more and more see-through. Eat an apple every morning and the apple becomes less and less tasteable. When we first arrive here, on the funny twin, things are hilariously tasteable, hilariously opaque; hilariously Hungarian, cerulean, cetacean, Dalmatian, peninsular, avuncular, vehicular, pink. But the years can really do a number on things. Reiteration renders things unhilarious.

Even summer can lose its quiddity, by September, when the land's goose is cooked and the geese are tired of waiting for fall and are flying away. The rest of us stay and contrive diversions, like follow-the-leader. Through time, leaders have been known as the Hammer, the Absolute, the Unavoidable, the Unready, the Impaler, the Quarreler, the Posthumous, and the Cabbage. Enough follow-the-leader begets in us an antipathy to both leading and following. Enough amassment of followers and the followers start to think, *Stop amassing me! I don't want to be amassed anymore!* They begin to realize that most leaders aren't even going anywhere, anyway, and shouldn't the primary requirement for a leader be to *go somewhere*? Otherwise the millions of followers are just a-standing there. The body will notice, even if the mind does not. The body wants to be a-zoom.

But the weather on Earth never gets as stuck as the weather on Venus. The poltergeist resumes: far-off streamers appear in the sky, with attendant rumbles, and rain arrives, and with the rain, the smell of rain, the petrichor, better than incense—but only when the rain is new, when drops of water splash on dry rocks. The petrichor—*this* is something we would follow, this is where we'd take off running, over stony soil scattered with sagebrush and wizened cactuses, through the Fossil Forest, past a pronghorn nursery, scrambling down the Sheepeater Cliffs, trying to keep up with the new-rain smell, hurtling north through abandoned Saskatchewan towns—but of course they'd be running, too—running south over saddle-shaped hills, through the clattery plants, perhaps stumbling into a new hemisphere, even, and down to where the plains turn into llanos and the sheep turn into llamas, where, if the runners lose track of the rainclouds and look up at the night sky, they might see a starry sea monster, a triangular guy, an otherworld or two. Though they have lost the scent, though they are as lost as burros, they get to see the smallest constellation, the Crux.

It is that long and dreadful dryness that reinvests the rain with such a charge. Repeated rain just smells like mud. Simeon sang his ecstatic song because he had waited all his life, till he was old and blind, to hold that one baby. But in the maternity ward, the nurse handling so many infants per day might not feel so ecstatic. Deprivation recharges ecstasy; this is true for humans, though not as true for nonhumans. Labrador retrievers don't have to be deprived before they get a charge out of something. For them, reiteration is like nonreiteration. The tenth peanut-butter biscuit is like the first peanut-butter biscuit is like the new snow is like the old snow is like the spring mud is like laundry day is like you getting up from the couch! Everything is hilariously mundane; every smell is like the petrichor, and everybody is like Emily Brontë, who has never been reiterated. Everything is like that undivestible melody, "Wachet Auf," Sleepers Awake. Sleepers awake, have another apple! There is nothing like an Earth apple.

THE EXPERIMENT

When two mathematicians duel in the dark, somebody is going to lose a nose. After losing his original one in a duel, Tycho Brahe sported a nose of gold. But it wasn't just his nose that was golden: when they exhumed his body they found there was gold in his eyebrows, too, and gold in his beard. It is easy to decorate yourself with a golden nose or a golden necklace, but growing a golden beard is something else. To grow a golden beard you have to have the gold inside, like fate, like something you can't take off at night. Of course, getting the gold inside is not all that hard—it is similar to getting the Ho Hos inside. You can get both inside if you sprinkle your Ho Hos with flecks of gold.

Other things are more difficult to get inside, like the

Sanskrit, or the weird. It's easy to decorate yourself with weird, to be weird but not deeply so, with everything outside the skull all seditious-looking but everything inside the skull highly normal. Anyone can have trippy hair but not everyone can compose Dvořák's Piano Quartet in E-flat Major.

There is another element that seems the sublimest element of all. It is golder than gold and harder to conjure; it is the thing that makes somebody go from holding your hand to really holding your hand. And the king who possesses it gets down on the floor and draws dolphins with the children. And although trying to produce it can feel alchemical, quixotic, like those medievals with their recipes for transmuting lead into gold, it seems too good not to try for, and therefore I have sketched out some experiments for attempting to engender in a subject that exquisite, elusive quality of humility.

Experiment One:
Disappoint the subject

Give him wishes and make sure none of them come true. Have him assume that life has big plans for him, and then let him discover that life was only shining him on. Have him expect to be hailed by the public but get hailed only by hail. Have him think he'll ride first-class but send him down to steerage. At anything he attempts, let him be middling at best. Feed him, all his life, small potatoes.

Experiment Two:
Feed him potatoes of any size

Since potatoes are known to be humble, perhaps enough of them added to the subject's diet would transmit that quality to his own composition.

Experiment Three:
Feed him psychoactive cactuses

Psychoactive potions hush the clamant ego and make you feel one with the grassy grass, the starry stars, the foggy fogs, the hoggy hogs. (And if you have anything accidental to say, you will say it. And if you have any stairs to fall down, you will fall down them. Tycho Brahe's moose, who trotted alongside him on carriage rides, preferred beer to water till one night he drank so much beer he fell down the stairs and thenceforth had no preferences.)

Experiment Four:
Reject the subject

If the right method of rejection is used, the subject may end up with humility and some tomato plants. When folks threw rotten tomatoes at your head, and you ducked, did you not find behind you, several months later, tomatoes growing like mad?

Experiment Five:
Give him a massive weakness

Chain the subject to a devastating fault, like a giant rock, like sloppiness or greediness or angriness, so life can chew out his insides every day. Actually this experiment has already been performed on every human ever. All ye embryos out there: you too will receive a rock.

Experiment Six: Have the subject try to do something impossible

Have her try to compose Dvořák's Piano Quartet in E-flat Major, or have her try to raise unraisable children. Some children are amenable to one *particular* child-raising method—the Filicetti method or the Dazdrazhinka method or the Gribble-Busby-Hapsell method. Some children are amenable to every single child-raising method ever conceived, the hooey ones included. And some children are simply not amenable to being raised.

Experiment Seven: Have the subject do something possible

Laundry is a possible project one can do every day that seems like it might eventually produce humility. Popping wheelies is also possible, but moderately possible, whereas doing laundry is extremely possible, up there with peeling potatoes.

Experiment Eight: Hit the subject with an asteroid

This would do a number on the subject's ego. However, it would also do a number on the subject, as well as her vicinity. There would be no more subject and no more vicinity. When it comes to asteroids, everybody's name is Porcelain.

Experiment Nine: Send the subject to space

If getting hit by a space object does not create the Beautiful Quality in the subject, we could try sending him to space, to go out there and realize how little he is and that he himself did not hang the moon. (Actually this experiment has never worked on me. I have never felt belittled by the universe: every time I go to space I come back raving about myself. Compared with all that dusty rocky nothingness I'm like a *rhinoceros.*)

Experiment Ten: Dissolve the subject

Let the subject be dissolved into that great solvent, society. Drop him in, as into turpentine, and watch him become same, samer, samest. No longer will he be personally proud. (Or personally anything.)

Experiment Eleven: Add dirt

The word "humility" is related to the Latin word for dirt, *humus*, as is the word "human." A dirty human is dirty dirt, and a dirty humble human is dirty dirty dirt. To increase both the humility and the humanness of the human, increase her dirtiness. Have the conscious dirt called Donna work with the unconscious dirt called Dirt. Have her try to get an eggplant out of the ground.

In the Middle Ages, some gardeners would plant seeds of gold, in hopes they would bring forth fruits of gold. At the beginning of the summer, Donna with her Eggplant Dreams might laugh at the silly medievals. But by the end of the summer, Donna's eggplants are still just dreams and she is very dirty and Dirty Donna is no longer scoffing at Medieval Millicent.

Experiment Twelve: Add years

This one is hands-off—and if it works, then all the other experiments are unneeded and we can dispense with the cactuses and asteroids and laundry and society. See if time alone will do the job: let the subject spend her life as she pleases—dithering, doing nothing, doing Spain, doing space. Will enough years grant her humility, or will they just make the subject into a wrinkly subject? The beauty of this experiment is that years are free, but the drawback is that they run out. Even the finest-equipped, best-funded laboratories run out of years. Infinity is what our science lacks, and our books, and our festivals and farms and everything else we do, even our worrying. You cannot worry for aye, anymore than you can be a kangaroo for aye. All the experiments have to stop, everything we do being finally untenable. (Thanks for the eons, said the flash in the pan.)

And things that matter matter on, and things that natter natter on. And if we did have infinity at hand, wouldn't we lack its opposite, and what would a book or a song or a kangaroo be without finitude? Isn't it their impermanence that gives kangaroos that je ne sais quoi? Doesn't our transitory condition lend an endearing quality to almost anything we do? Like the guy down the hall who's always measuring his erudition—last year it weighed forty-eight pounds seven ounces, this year it weighs fifteen pounds fourteen ounces. Insufferable, if he were infinite—but when you factor in his finitude, his fixation is kind of sweet. Who would begrudge the white mouse a little sashaying around her laboratory cage? Why resent the pig who thinks he's one of the great pigs of the world? Maybe he *is* one of the great pigs of the world.

Still, we have known a few characters who possess a real, ravishing humility, who even without the help of infinity or impossibility have all the hauteur of puppies. Who, when you run around the house frantically looking for your wallet, scramble after you into every room, or, if they are old and ailing, follow you with the beams of their rheumy brown eyes. Who, when you sweep them off their paws, plying them with goodies and treats and runs in the park, thrill to the fundament of their hearts. For you are the one, the very one, yes you are the one who hung the moon. And they will eat potato-peel pie, moldy-old pie, dirty-socks pie, all the varieties of humble pie. And because they are not trying to win fans, or customers, or acclaim—because there are none of those impairments to their experience of the world—when they chase the ball they really chase the ball, and when they run across the field to greet you they really run across the field to greet you, and then they really jump up and really knock you down and really lick you to death, humility being the extravagant thing that it is.

HOW TO COUNT LIKE A PRO

Good morning, animals, and welcome to the first lecture in our series "Jumpstarting Your Career at Earth Inc." Can we offer you some puddles to drink? Please settle down, quit flapping, stop bellowing, retract your claws and lower your tails, thank you. Now, before we start edifying you, we're going to level with you. We've been concerned about your performance lately, which has been, well, less than enterprising. You just do not seem very plugged-in. Of course, we understand that you are all anachronisms, and that as anachronisms you have been "grandfathered in." We are not expecting you to become astroanimals or anything; but out of concern for your viability, we'd like to help you salvage your sagging careers and regain some relevance. This is an opportunity to hitch your wagon to a star! Humans are stars, having risen through the echelons of Earth to practically transcend it! If you attend our seminars, we will not microwave all of you, and we can help you get on the same professional page as us.

We used to be on different pages ourselves, or sometimes not even on a page at all. Our ancient calendar had only ten pages—ten months, starting with March and ending in December, which left winter just a numberless stretch of days. But since we reformed the calendar, adding January and February, there are no longer any off-calendar days. All days are on-calendar. Time being money, it is financially foolish to leave even one day off the calendar. Modernity means business!

The calendar reformation is a good example of what is possible—you can reinvent calendars, you can reinvent *yourselves*. We hear some of you going around saying, "It is my lot to be a yak," or "It is my lot to be a mudpuppy," or "It is my lot to be a green water dragon," or "It is my lot to be a bagworm." Do you not realize how fatalistic you sound? You sound like turnips! In this series of lectures, we are going to set out some important principles to help you break out of "turnip thinking."

To begin with, let's discuss the power of numbers. As the father of eugenics wrote, "Whenever you can, count." Larry here will now hand out stickers with this motto on them, for you to post on your burrows and bowers, the sides of your nests, the entrance to your caves. The counting habit is going to help you cultivate the three R's—rationality, reasonableness, and regulation—you with your unregulated ids—and eliminate subjectivity. Subjectivity is like a banshee, nonexistent and therefore easy to eliminate. Sometimes, when you see the emerald and ruby and sapphire sparkles on the snow, it *seems* like you are rich; sometimes it seems you can't get along without someone, seems winter will never end, seems the moon is abnormally big coming over the mountains. But measurement dispenses with seeming: the bank account is low, the moon is normal-sized, etc.

Count whenever you can! We use our fingers and toes, but you can use your toes and toes, or pincers or flippers or whatever, and you snakes can make toothmarks in a branch. However you do it, you should start counting everything you see—clods, clams, azaleas, skunks. Plants, being stationary, are generally easier to count than animals, but stay away from furze, because it has no plural. "Furze" is one of those uncountable nouns like "information" or "butter." When you do start counting animals, start by counting strangers—and remember not to look into their eyes, lest they lose countability.

Counting strangers is like counting words in a foreign language. If someone writes to us in Kickapoo—"Ămănutci wīpăni," or "Măgānăguhanu, nezegwize, ähitci īna Wīza'kä'a"—counting the words comes easy. But if somebody writes to us in English—"I need you," or "I don't need you," or "Let's get revenge on the old buzzard"—we can get caught up in the meaning and forget to count. Meaning undermines objectivity.

The words that really matter are the words for numbers—one, two, three, four, five—or yan, tan, tether, mether, pip—or hant, tant, tothery, forthery, fant—but see there, how arbitrary words are! Someday we should replace all our words with numerals. Numerals are absolute, and think of their stamina compared with words. Numerals never run out—you just add one more, one more, one more: sheep 4, sheep 5, sheep 6. You don't have to come up with a new name for every sheep, just a new sheep for every number, and if sheep 7 gets squished, another one can take her place. Names are not as transposable, and names can sound like children's songs—Waxahachie, pumpernickel, the village of Nobbin, the hill of Nabbingo.

Names are also unnecessarily meaningful, like the old Armenian words for the days of the month. They had a day called Tumultuous and one called Hermit and one called Dispersion, and a day called Beginning, which came right before the day called Beginningless. How today could be Beginningless when yesterday was Beginning is a boggling question dispelled by calling the days Sixteen and Seventeen.

Now, we humans have the advantage that our contemporary culture is principally composed of countable things. Sports and politics and business and social media, with their rankings and followers and prices and indexes and polls and points, help us keep our heads thoroughly in the numbers—unlike you giraffes with your heads in the clouds. We do our best to quantify the clouds, but they and other components of the weather elude us sometimes. Lightning is elusive, incendiary. (Literature used to be like lightning, but now that we have subjected it to Big Data, literature is more like sheep. Never have we burst into flames when we got hit by a sheep.)

Anyhow, if ever we get flummoxed by the weather, we can always turn to our clocks. Clocks are the consummate counters, even better than the bankers because they never sleep and especially they never dream. No minute is off-clock. What we know about ourselves, from research articles we've read, is that what we find most attractive in a face is symmetry. It was inevitable, then, that we would fall so hard for clocks. We have actually entered into an exclusive relationship with them and can't imagine being tempted by someone less symmetrical.

To be a great counter, like a clock, one must be on guard against perceiving distinction—but, of course, where there are no distinctions, none shall be perceived. You animals have your own advantage, in that within your species you all have the same faces, like nickels, so you probably don't even need to worry about sticking to strangers when you count! Your relatives look exactly like your strangers, and it's not like you hamsters would be counting hamsters without a hitch but then your dream hamster scurries by, making you lose count. This unfortunately does happen to us sometimes, though not with dream hamsters. The worst is when we've been counting people with clocklike consistency but then the *granddaughters* run by. Granddaughters mess up the metric, and they grapple us to them with their little fingers. This is called the adoration problem, and it contaminates the purity of calculation.

The point is to never be snagged by the particular. "Everybody to count for one, nobody for more than one," wrote Jeremy Bentham, and this means that everybody is everybody. Everybody says, "I'm not everybody," but of course everybody *is*. Just listen to two everybodies arguing about who is everybodier: it is as ridiculous as two twenties arguing over who is twentier.

We feel sorry for those in the adoration business, the *adorazzi*, the mystics and musicians, muddleheads when it comes to numbers. "Better is one day in your courts than a thousand elsewhere"—the Psalmist's singling out of one day like that is so delusional. There are no deluxe days, just as there are no deluxe hamsters. Every day is every day; every day is precisely the same little square inch as every other day. The Psalms are full of bad math and seemings and vicissitudes. There is nothing like a musical instrument for exacerbating vicissitudes, and in the Psalms you find lyred people, luted people, fluted people, tambourined people, completely abandoned to their vicissitudes, and harebrained harpist kings longing for someone *invisible*. (Bad enough to crush on someone visible.) We think of the Psalms—actually, the whole Bible—with dismay. Here we are computing on our computers, and there is Miriam getting all carried away on the timbrel. *Still playing the timbrel, Miriam?*—that question really encapsulates our thoughts about old testaments.

The one distinction really valuable to make is the one between Essential and Superfluous. Up to a certain number, things are Essential, and over that they are Superfluous. As names go, Superfluous is a good one to have on hand. Any offspring beyond two or three you can name Superfluous. Having counted ninety-nine of his sheep into the fold, the sensible shepherd will call the hundredth one Superfluous and turn in for the night. No need to traipse around in the chilly, rainy, brambly dark searching for a lamb called Superfluous. Counting enables one to distinguish between Sufficient and Surplus, although somehow in India the cows got into the sacred racket, which makes no such distinction. (Indian math has a history of being irrational—India is where they came up with irrational numbers and mathematical infinity.)

When we were two we loved all the cows, and every hamster was our dream hamster. But then we grew into good calculators. Calculation converts the sacredest things into inventory, the cows into mooing merchandise. Not all cows are in superior condition, but we can still use their umbles, numbles, hides, and hooves. Not all wines turn out wonderfully, but we can still serve the cheap ones after the guests are plastered.

To use the usable, squeeze the squeezable, skin the skinnable, drink the—hang on, come to think of it, you leeches seem to know this principle already! We've never seen *you* in a tasting room, swishing blood around your suckers, spitting it out into a bucket, getting all choosy, all highfalutin about "mouthfeel" and "fruity notes." Blood is blood and everybody is a decanter.

Success being measurable by numbers, we have to

admit we humans are outnumbered, and thus outsucceeded, by another kind of bloodsucker. For every one of *us* there are maybe ten million of *them*. They are ten million times as efficient as we are, ten million times as effective—and this point brings us, finally, to an exciting announcement. The next presentation in our series will feature the triumphal story of these super-successful figures! Everybody be sure to bring your cavemates, burrowmates, warrenmates, and puddlemates to the afternoon lecture on Thursday, entitled "The Growth Mind-Set," where we will all get to hear an inspiring message from the mosquitoes.

STRANGERS

Omniscience is a big responsibility. Now that our knowledge, having accumulated for eons, is almost complete, we are not taking our responsibilities lightly. We try to be tireless purveyors of information, faithful disseminators of facts, avid openers of eyes. Our information has now reached everyone in the world except all the babies and a few schismatics on Raspberry Island. Babies are such dilettantes. Modern babies are no better than ancient babies, but as soon as they buckle down and learn our language we will be able to update them. Sadly, the schismatics are too busy marching in holy clockwise circles to listen to us.

There is another rather large group impervious to

our information: the animals. Gladly would we bring them up to date, gladly would we clue them in—better a clued-in wombat than a clueless wombat, is what we say. We've even tried assembling animals with some seeming brainpower (not anemones), animals who'd be interested in bettering themselves (not panda bears). We did not want to drown them with a flood of knowledge, all at once, so we planned short seminars with titles like "How to Count Like a Pro" and "Technology: Your Portal to Reality" and "Your Dendrites and You." We set it up in the open air, with slides.

We had thought that, in response to our presentation, the animals would howl in gratitude, as if to say, *Oh thank you, you explain like an angel! We feel like we are waking up! Oh speak, speak on!* But before we were even five minutes into our lecture, the magpies had flapped off and the tortoises were trundling away and the pigs were just looking at us like we were blowing our noses. At first, when some processionary caterpillars came down from their silken tent in the oak tree, we thought they were going to take the opportunity to become better informed, but then they just processed on by to get to another, leafier tree. Preaching to the animals is a discouraging business. It's like: lose your lasso, lose your audience. It's like they don't even care about being well informed.

Educational disparity can be detrimental to relationships. The intellectual will necessarily outgrow her country cousins and have to distance herself from them, unless they have some kind of cachet to compensate for their backward thinking. But no kangaroo is a kangaroo of consequence, and all the geese are gauche. We just cannot include the uninformed opinions of echidnas and marmots in our decisions. Actually, we are not sure if marmots have opinions at all, since opinions, once they possess a body, propel it along a straight line, torpedo-like, and marmots are meandery. (Sharks have opinions for sure.) Anyhow, neither can we include their uninformed nonopinions. Animals these days just constitute an enormous, ill-informed, infantile rabble, invulnerable to facts, though vulnerable to just about everything else.

Balder was the opposite. Balder the beautiful god was *invulnerable* to everything. After he dreamed that he would die, his mother, Frigg, made everything in the world pledge never to hurt him. Beeches wouldn't fall and hatchets wouldn't slip and hornets wouldn't sting; bulls would keep their cool and privet shrubs wouldn't tempt him with their murderous little berries and eagles wouldn't drop a turtle on his head. Time promised not to mangle him, rivers promised to slow down, his friends promised to never grow tired of him.

Even pebblesnails and pattypan squashes turned in the required signature—everything except for one glossy-green little plant. It's not that mistletoe is treacherous, or lawless, or even rascally, really; it's just that mistletoe can be remiss. Leave it to mistletoe to forget to turn in paperwork. You can work out the rest of the story: *Balder the Beautiful is dead, is dead.* (If you brush against a mistletoe, you can still hear it whispering, *Mistletoe is sorry, is sorry.*)

But animals have no mother Frigg, just mother animals. No mother animal has ever been able to collect even one signature from the world. Animals are like Balder in a world full of mistletoe. They are Balder backward, vulnerable to everything, except one thing: *information.* How is it that they are so unresponsive to information, when they will respond to music, even the homeliest music? There goes Anne Murray on the radio, singing her slushy songs, and the mouse who should be scurrying away starts to sway back and forth on the armrest of the couch. We have seen Pomeranians under the rumbling piano calm down like King Saul. Oh sure, we listen to peppy music when we're doing the dishes, but we know that music's main function is to prepare the brain for math. Music is John the Baptist and math is the Savior of the World. But for all their receptiveness to music, animals have a tin ear for math.

Supposedly they used to listen to that little Francisco who preached to them in the medieval forest. He's been called presumptuous for preaching to birds, but birds, of all people, are not going to stick around if they are unimpressed. We just find it sad that they *were* impressed, because Saint Francis was full of beans. If he was the last person the animals listened to, then they must still believe that hogwash about the moon being their sister and the sun their brother, about people and fishes and burros and sparrows all being brothers and sisters; that none of us deserves credit for our own loveliness. Such misinformation pains us to no end.

Now, of course animals have relatives; the genome tells us so. Many sea snails belong to the genus *Ittibittium*; giraffes and okapis are cousins, with their ossicones and prehensile tongues. Where there is genomic overlap, there is physiological resemblance; thus you have your uncle's ankles. But we suspect Saint Francis was speaking less physiologically than familiarly. As if we could be on a family footing with pigeons, those dirty burghers, or aardvarks, or geezer monkeys. (Though we do *not* mind the idea of being related to lusty monkeys.) As if we could be related to bovines, so blasé, or weasels, so unpatriotic, or army ants, that blind collective, or shrimp, so snacklike. As if we could be related to giraffes, who keep their thoughts to themselves, or hummingbirds, who take themselves so lightly.

As if the sparrow could be our brother, so common and dun, so negligible, spending all that energy sustaining his negligible nestlings, propagating what, in the world, but negligibility? Or the strawberry frog, who mothers in the most ridiculously effortful way, producing only three tadpoles and then carrying each one to a different bromeliad-puddle in the forest, tending, over the next weeks, to each of them individually, making eggs for them to eat and then serving them up, like a tiny red waitress in a twenty-four-hour café with the tables acres apart. Instead of having thousands of tadpoles and leaving them in the hands of Statistics, the strawberry frog is mathematically fatuous, laboring all day for her three children as if they were worth a lot, whereas actually each froglet is only worth thirty dollars.

As if the crabeater-seal pup in Antarctica could be our baby brother, galumphing up the Dry Valley after getting separated from his mother. He needs to get to the sea but is headed the wrong way, bumping clumsily over sharp rocks in the katabatic winds. Instead of going north to the ocean, he travels souther and souther, getting so starved along the way he starts to eat gravel. He passes other baby seals, mummified for millennia, with their mummified tummies full of gravel. And if you try to turn him around, he *bites* you with his sharp white baby teeth. Being righter and righter ourselves, we find it impossible to relate to someone wronger and wronger.

Tellingly, the seal pup shambling to grief has no name. Siblings always have names; but imagine sea squirts named Opal and Mario and Delphine, or whiskered bats named Stu and Stanley, or a duckling named Richard P. Duckling. Gnus might be more distinct if they were called something other than Gnu, Gnu, Gnu, and Gnu.

Another trait of siblings—unlike strangers—is that their anatomy is not the most significant thing about them. Something would seem off about the fellow who, when asked about his sister, said, "Well, Gladys, or as I prefer to call her, Homo Sapiens, has four incisors in each jaw and helical outer ears and a slight asymmetry in the placement of her eyes." Because they were sniffly little kids together, because he held her head above the bathwater once when she was a baby, because they shook the tree branches to make snow fall on each other, Gladys is somewhat transparent to him. He can see through her head to her mood, whether she is frantically sad or wistfully sad, whether something she is not saying is interfering with what she is saying. Whereas strangers are all pretty much opaque. Animals are opaque like strangers; even jellyfish and glass frogs are opaque like bears; and when their anatomy is gone, they are gone. This is not true of brothers and sisters.

Such animal opacity accounts for the last remaining deficit in our knowledge. Total omniscience would mean understanding every last creature on Earth, every individual animal, every muddled, modest mind of "the least of these." Yet this deficiency is presently shrinking, as the number of animals is shrinking. Fate is going around with a stick, knocking the birds out of the sky. Frogs of every stripe are going belly-up. Monk seals and albatrosses and vaquitas and manatees are plunging to the bottom of the sea, in some kind of sinking race.

No schism occurs without a few pangs and regrets, of course. The animals were heretics, but sometimes they seemed herer than we were. Maybe it was because bits and bytes flew over their heads. Maybe the animals were only ever going to listen to somebody who spoke in moons and stars; but we never would have stooped to that, pandering to the animals. If only there had been some river that could have washed all that magic off, all that moonshine, without washing them away as well.

THE CATASTROPHE

For a long time, about a billion years, green and purple photosynthetic bacteria had the run of the planet. They floated in community mats, absorbing light and engaging in carbon fixation. The atmosphere was stable, the environment unchanging: that period of quiet stability is called the Boring Billion. Then there was a Great Oxygenation Event, which ushered in the next era, called the Interesting Billions. Squeaky species appeared, and quacky species, and screechy species, and one very speechy species. Hadrosaurs appeared and crashed around, monkeys appeared and fell out of trees, snakes appeared and freaked everybody out. Wolves appeared and turned into poodles, wolverines appeared and devoured their betters and their biggers.

Betsy the fiddler appeared and played "Go to the Devil and Shake Yourself." Now, it is hard to know much about bacterial cogitation, bacteria being so sphinx-like, but maybe every once in a while they think wistfully back to that earlier era, before all the hullabaloo. Maybe they wrinkle their noses at us from their sludge lagoons.

But of course, if they don't like the commotion, then they should never have produced all that oxygen—for that is what they were doing, during the Boring Billion. Well, isn't that just the way it goes? There you've been, meditating tranquilly, attaining such absorbing focus on your inhales and exhales that you have no idea how many epochs have passed—only to look up one day and see that your respiration has been sustaining hippos and hippies and tropical boubous, people singing "El Loco Cha Cha Cha," wallabies going boing-boing-boing. The Great Oxygenation Event is also known as the Oxygen Catastrophe.

And ever since the advent of oxygen, nature has been on one long march to imperfection—except for the popes, of course, who have been marching in the opposite direction. Over the centuries, the popes experienced something called "creeping infallibility," but for the rest of us it has been a story of creeping fallibility. The unicellular organisms just do not seem to make as many mistakes as us multicellular individuals. The more innards, the more errors. It is even true for plants. A vascular plant might mistake a warm winter's day for spring and send the sap flowing prematurely, but liverworts would never do that. To err is maple, to err is celery. Panegyrics are passé. Humans do not have the corner on fallibility, though we might have the corner on run-on sentences.

To err is manatee. A manatee may mistake a swimmer's long hair for shoal grass and start munching away, oblivious to the attached figure. To err is baby elephant, tripping over her trunk. To err is egg-eater and moonrat and spaghetti eel, and whales, who eat sweatpants. Even the dinky species mess up, like the gerbil father eating his own babies like ice-cream cones.

To err is human: we make mistakes after which we get the early-morning sorries, and mistakes after which we get no more mornings. Look at our inventions—to err is airplane. Look at our conventions, which seem to be standardized mistakes. Listen to our musical mistakes—all those piano crimes, clarinet crimes, and drum crimes being committed on stages all over the world, maybe in one's very own apartment building.

We make conscious mistakes and unconscious mistakes, incompetent mistakes and competent mistakes, wild mistakes and timid mistakes, although with timid mistakes timidity is usually the main mistake. Timidity turns you into a personless person, and a personless person is no better than a gerbilless gerbil. Oleg minus a leg equals Oleg; but Oleg minus Oleg equals zero. Stop phantoming around before your time.

To err is human, yes—however, since humans like to identify as all kinds of different things, the adage can be made more specific. To err is grammarian, garbage collector, saint—Saint Augustine sure had some erroneous ideas there. To err is penny-pincher, to err is pharmacist. There seem to be as many different kinds of mistakes as there are identities, even angel mistakes—remember that one time when you had to fire your angel?

Those who identify as sea slugs make mistakes, too. To err is sea slug. However, to learn from your errors is also sea slug. If you attack a flabellina and it stings you, you can undulate away and remember, the next time you see a flabellina, that flabellinas are the nastiest, most maleficent little poisony pink-and-purple fringes, and you can prudently not attack it. From sea slugs we can learn to learn from our mistakes. (From flabellinas we can learn to sting our attackers.)

Once, my borscht tasted strange, almost soapy, and then a pale-green wax congealed on the surface, and a wick floated up, and thence I learned not to balance sandalwood candles on the ledge by the stove. One night I forgot to put the lid on the honey jar, and the next morning I found a mouse swimming in the honey in slow motion. Once I invited Gwendolyn and Gwendolyn to the birthday party but forgot to invite Gwendolyn. These and millions of other blunders have taught me millions of things. The Earth is a good school.

But of course we haven't learned from all our mistakes yet. We keep throwing all those sweatpants into the ocean and giving our hearts to telephones and televangelists. We keep celebrating the celebrities while turning the baby cows into schnitzel; letting our herds overgraze the grasslands, rendering the land so dry the only water is tears. We keep getting into a torrid affair with Dogma, leaving Laughter and Philosophy to huddle together in the closet like two neglected children.

But if we could retract all the mistakes, all the soggy sweatpants from the ocean, all the soggy words, fishy words, we'd said, all the musical crimes—if the mistakes were all retractable and had hinges and could fold up and recede, I suppose we would have to fold up and disappear too, for even if we were not ourselves mistakes, some baby in our lineage surely was. The fallible wallabies would have to go, along with the other fallible furbrains and featherbrains. Goodbye, wallabies; goodbye, vixens; goodbye, emus and buntings and howler monkeys; goodbye, catastrophe.

Absent thee and absent me, the world would return to serenity, to the longueurs of perfection, and there would just be the bacteria and the popes left. The only problem with retracting everything imperfect would be the goodbyes, for we would have to retract all of them, for there has never been a perfect goodbye, not one, and goodbyes can prove impossible to retract.

COUNTERSHINE

Ever since leaving the anonymity of the womb, we've been questioned about our identity. Some questions are easy to answer, like "Are you a man or a mouse?" or "Are you a private or a generalissimo?" Other questions are trickier, such as "Are you an optimist or a pessimist?" or "Are you a materialist or an immaterialist?" Feeling something between optimism and pessimism, you might be more of a pessimop or an optimess. And "materialist" as an identity seems so general, because matter is so general. You might be into solids rather than gases, or you might prefer liquids, solids seeming superficial. Oh sure, land has a pretty face—all bluffy and wheaten and green in the gullies. But that's as far as you can see: land you look *at*, whereas water you look *into*. Water is deep, like lemonade.

Even if solids *are* your thing, you might not be into all of them. You might be into doilies but not bellows or bellows but not bauxite or bauxite but not chocolate or chocolate but not cellos or cellos but not cellophane or cellophane but not jalopies or jalopies but not jalapeño poppers or jalopies and jalapeño poppers and bauxite and foxy bangles and lilac-scented floating candles but not dreadnought battleships; or you might be into all those things but not nanobots.

Of course, there are distinctions when it comes to the *immaterial*, too—like you might be into time and gravity but not augury or angels. Or you might be into some angels, like the six-winged amber ones, but not the messenger of death. Since there are so many variables, since we are always being convinced by disparate things—the grim and the breezy, the bitter and the sweet, the opera and the honky-tonk—do we have to commit ourselves to just one identity? Do we have to be either Immaterialism Irma or Materialism Mack ("Buffaloes ain't got no spirits")? Since we are always being bundled from day into night and night into day, and since night and day are always showing us different things, persuading us to different ideas—why can't we just shift our identities with the shifting light? Why do we have to join a camp? Isn't the Earth camp enough? Our planet has a lot of ambience, and they say its ambience came from within, that the Earth emitted its own variable atmosphere. Why not vary, too? Why not be materialists by day and immaterialists by night?

So one thing I like about matter is that I don't have to activate my third eye to apprehend it. I don't have to chant "Om" 108 times, collaborate with amethysts, do the flying-pigeon pose, and take zeolite supplements to be able to see monster trucks. To see graffiti or crab apples or late bloomers like the saffron crocus, or that castle-looking island in the deep blue lake. To see the red in the rainbow and the rainbow in the trout and the trout in the pond. To see how a tree bleeds sap when its branches are removed (a tree's branches are removable, but then it bleeds). How the needle on my compass starts to whirl around when I stand at the North Pole. How dried lettuce sparkles when it burns. They say the angels are jealous of us down here, and if I think about it, I am jealous of me too, getting to see all that graffiti and ride in all those trucks and wear all those sweaters.

That blue lid of light over us during the day is a *particularizer*. When they take the lid off at night and we can see for only trillions of miles, it is easy to *generalize* about the materials down here, the weeds and waves and people in their fifties. When you can't see things, it is tempting to imagine them identical—the mind is such a mimeograph. It is especially tempting to generalize about generalissimos, but then daylight brings back their particularity, and you see how one generalissimo has wavy white hair and one generalissimo plays the ukulele and another one, in exile on a tiny island, wears all of his shiny clinky medals even when he's napping.

During the day you can see materials in interesting combinations—bear and jogger, bear and butter, zucchini and sunshine, the porcupine and his inamorata. The bear and the jogger collided and rolled off the trail together and then, discombobulated, ran away from each other; another bear climbed into a cabin through the window and binged on the butter, and now, with her buttery paws, is having a hard time opening the sliding door. The sun has trained up the zucchini in the way he should go, and the zucchini is profitably productive. (If you asked some successful zucchinis what their main influence had been, they would probably say sunshine.)

The porcupine's relationship with his inamorata results in three fuzzy little reddish porcupettes. The bees' relationship with thousands of asters results in one laboriously composed spoonful of honey, which, if they had their druthers, they would keep for themselves. Ants, if they had their druthers, would have you *not place their log on the campfire.* If you do, they will hurry out, carrying their squishy white babies, depositing them in safety, then going back to retrieve more and more of them from the burning log, until they are wobbly from the smoke and bringing out fried babies.

Sometimes matter combines in such a way that we stop being jealous of ourselves. If the five-year-old in your family has not been putting his books or cars or blocks away, or the drawings of winsome monsters, or the blue sheepo (half sheep, half hippo) he pulls around in a wagon, and the three-year-old has not been sorting through the mail or washing the dishes, and the baby hasn't been sweeping the crumby floor, you can either do it all yourself or just wait till the sun goes down and clutter disappears and your house might be a shipshape house full of utterly elegant, utterly unbroken things.

As good as consciousness is, unconsciousness is commendable too, because when people are asleep there is no behavior. For all the consciousness-raising campaigns out there, it seems like there should be some unconsciousness-raising campaigns too. When everyone around you is unconscious, then your mind can hie to the moon, hie to Ohio, hie to people who no longer have any whereabouts, and sweet old dogs. Though it is not possible to memorize a dog like you memorize a poem, still there is something to hie to after she is gone.

Then, in the dark, is when it seems there are things in the wings. As darkness filters out the visible, so it features the *invisible*—that which is not obvious or concrete or ironclad, which will not be clad in any metal; which is not truck, or sweater, or medal, or all the rage; which can't be hawked or haggled over or requisitioned, and can't be cracked. It can't be cracked because there is no code.

And words might come to mind, such as "The rain it raineth every day," or "The river Jordan is chilly and cold," or "O Maurizio!" or words with a little blue beat, or astral words spinning around. Words are an immaterial material: artists who work with material materials have to save up so they can buy their paints and clays and bronzes and woods, but words are free. Not that words cost nothing—how many stars must Shakespeare have visited to compose his lines?—but they don't cost dollars. And when his words come to you, in the middle of the night, it's like the bees are personally bringing drops of honey to your lips.

Of course, it is possible to tire of the immaterial, to fall out with abstraction. The solitude of darkness can be searing—solitude requiring no more than, but also no less than, one soul. At night your life can seem like a book about somebody who never shows up. And abstraction can of course be misused. We've all seen people getting into the abstraction racket, promising goodies out of their cabinets of nothing, claiming to have cracked the code, getting all pious and proprietary about nothing, which seems even more preposterous than being proprietary about the moon. Such abuse of abstraction can make you want to run headfirst into a brick wall, to beg for great chunks of concrete or wheelbarrows full of slag.

Toward dawn is an overlapping of night and day, when you can almost see and almost can't, and all of Animalia seems amphibious, ambiguous, the creatures funny junctions of the seen and the unseen, like origami with woes. A fencepost and a farmer can both get muddy, but only one of them can get discouraged; a compass and a pizza-delivery driver can both get disoriented, but only one of them can get dispirited. The dispiritable ones prove that material is not everything—form is not everything. Not that form is nothing, but just having a *Mohawk* like a zebra don't make you *cool* like a zebra.

During twilight, when night and day converge, is when you might see the Gegenschein. The Gegenschein, composed of dust particles five miles apart in space, shines faintly in the sky along the zodiacal belt, directly opposite the sun. It was always elusive, that countershine, even in the Age of Sea Lilies—sea lilies never having gotten it together and invented the incandescent bulb. But our modernization of the Earth has modernized the sky, too, rendering the Gegenschein generally invisible. There are some things that light can obscure.

Used to be, day and night had reciprocal rights to us, and after day had substantiated the seen, night would take its turn, substantiating the unseen. But in this Age of Lightbulbs, we can clamp down the lid of light all night long. We are like ants rushing out of the tormenting darkness, or zucchinis, with light our primary influence. Light substantiates the immediate—there is nothing in the wings, buffaloes ain't got no spirits. How clearsighted we are, how sober; how strange to think we ever sought, with these our removable bodies, that which cannot be removed.

THE WANDERER

It is customary, when one is reviewing a klezmer concert or a kabuki dance, to sit through the whole performance first. But there is this one extravaganza, already in production for five million years now, called "Earth." Because it is so full of redundancies, so repetitious in fishes and winters, we feel we have seen enough to get a handle on it; we would like to set out our critique of the planet's aesthetic merits and failures before we are toast like Tacitus. There was one critique, once, that said it was "very good," but that was affectionate and antediluvian; it is high time for a dispassionate reassessment of the Earth as art.

Two felicities we would like to commend, first off—the artist's facility and his versatility. There is an effortless, offhanded quality to the inventions here, as if sneezeworts and salamanders came easily to him, with no strain or torment. Of course, such extreme facility does bring up the question of *taste*. Let's just say that, if we were able to conjure anything out of the blue, it would not be a blobfish. As to the versatility on display, it is equally a virtuoso performance—viz., voodoo lilies, vireos, chiffchaffs, sapphires, walking leaves, surfing snails, hellions, moppets, yahoos. Such variation leaves us reeling, and also a little suspicious. To a certain degree, versatility is admirable—we admire someone who can speak Japanese and Hungarian as well as the business and boxing dialects of these languages. But there is such a thing as being overly versatile—if this person also speaks Grackle and Grampus and Baby Just Born, that starts to weird us out; she starts to seem shifty, promiscuous.

Imagination unchecked can result in a mishmash. There should be a common thread to all of an artist's works, a uniformity of purpose, a marshaling of ideas and characters. The alignment of a writer can be maddeningly hard to detect if her characters are free; we just think the world could have used a *teensy* bit more autocracy. It's not that we want the animals all to be square; there should simply be some consistency, unity, *something* tying everything together, to signal the values and priorities at work here. How hard can it be, if you can make porcupines and jellyfish out of the blue, to make uniforms for them to wear? Blue jackets, neat hats, chevrons for favor. As it is, the world seems deficient in uniformity and purpose. Granted, there are mini-purposes here and there, like how within his swarm a mortuary bee has a purpose—dragging away the dead bees. A mopper's place in Mattress World is clear, but what is the mopper's place in the universe, the universe being inscrutable? When everything is mad, even the exigencies are mad. Sweeping, mopping, shmoozing, morticianship.

We often feel that the artist is toying with us, being purposefully opaque, making us try to winkle out his meaning. Sometimes a cloud resolves into a camel, and then we think, *Aha, so* that's *what he's getting at.* But then we think, *But wait, we don't really know what camels are getting at.* The figurative art here is as enigmatic as the abstract art. Clouds resolve and dissolve, like camels, urchins, efts, the Inka Taky Trio. Here is something else we find confusing: unlike in a photograph with a celebrity in it, we are not sure where we're supposed to look. Are we supposed to look at the snow falling or the tree behind the falling snow or at the chickadee in the branches? A more accomplished artist would have *featured* certain characters, *exalted* them, inserted more focus, more signals, and—crucially—would have made the signals actually signal *something*. As it is, who knows what it means for the green hills to sprinkle with gold every spring, for the cranes to creak like that, for that one tree to remind us of our grandmother? In his way, Anonymous is as squirrelly as Shakespeare, who took perfectly good stories with coherent motives and clear trajectories and subtracted all that. The world seems similarly subtracted from, unscrewed, mystifying. Mystification is not an end. It's not that we've never been tempted by mystery, with his darkening, deepening eyes; it's that we can anticipate to what wordlessness such an assignation would lead. Our words are our sovereignty; we daren't yield to that punk, mystery. Mystery's a punk like Mendelssohn: songs should always have words.

Then there is all the flummery of the seas, as if spangles and flounces had fallen off their gowns and commenced to lead lives of their own, with consternations of their own. It is disturbing enough to look underwater and find a tchotchke. Violence we get, predation we get, heads getting bitten off—deep things should be abominable. Deep things should not be flossy and twee—little turquoise trinkets waggling around—and trinkets should never have worries. A worried trinket is aesthetically incongruous, like a flibbertigibbet at prayer. But seahorses do have worries and disappointments sometimes, having dropped their clutch of eggs into the seagrass, or having wrapped their tails around a holdfast that floats away—a plastic straw.

The wicked thing about punctuation is that it always has the last word. You might be reading a careful, sober, measured statement but find, at the end, upending all that sobriety, an exclamation mark! A sentence may seem to be tendering absolute verities, but then a question mark can come along and subvert it all? Seahorses are not just frivolous squiggles but tiny spiny question marks with a fan on their backs; append a seahorse to any of your precepts and see if the precept does not begin to wobble. Even a conviction capacious enough for a landhorse will hardly be able to accommodate a seahorse, for what conviction can accommodate a question mark? The seahorse is certainty's nemesis (albeit a nemesis who swims away from you, back down to her gorgonian sea fan).

Now, to address a general difficulty we have with the characters here: in ascertaining anyone's identity, we rely primarily on extraction. Someone's background, more than her patience or parsimony, identifies her, allows us to place her with her identicals, like a Buick with Buicks. But the players in this piece, finches and noddies and ouzels and froglets, carry no sign of their extraction, no air of whence they came—there is no whiff of the void, not a trace of nullity about them. Extracted from nothing, they arrive screaming, scrambling, wanton, wriggling, hailing from nullity with entity coming out of their ears. Entity defies identity. By entity, of course, we do not mean the material: material comes not from nothing but from material and is easily identified and handily reused, like a marzipan moose being reshaped into a marzipan mouse. No, entity is that which looks, or speaks, from out of the material, and startles us. (No color is so startling as clear.) Entity is always getting to us, entity's plights and entity's moos: there's a barn we don't care about, then we hear a moo. There's a hermitage we don't care about, then we hear a guffaw.

Certainly, with entity housed in such a profusion of forms, this can be fascinating stuff; the problem is that it is so overwhelmed by excess. We don't know if this is perversity on the part of the artist, or if he just got frog-happy, periwinkle-happy, and forgot his master design. If we could just establish the genre, whether this is supposed to be comedy or tragedy or romance or what, then, following the imperatives of that genre, we would retain only those elements that contribute to the work as a whole. Armadillos are never apt. Such effusions should obviously have been scrapped at the outset, being excessive whatever the genre. Butter-scotch or banana, no pudding should be over-egged. Here we have the needlessly pretty, the needlessly weird, the needlessly unpleasant: Natterjacks needn't smell like burning rubber. There are badgers too bilious, koalas too torpid, cockles too vacuous, broncos too bucking, orangutans too Rabelaisian, olms too ambiguous, beetles too countless (counting soothes us like brandy), Pomeranians too combustible, penguins too reticent, motmots too resplendent—such ridiculously tailed birds have to back out of their burrows like bustle-bottomed ladies backing out of a carriage.

Birds in general are too privileged. Privilege is not inspiring but alienating—think of the difference between telephony and telepathy. It is inspiring when somebody makes deft use of the telephone, alienating when she makes deft use of telepathy. It is inspiring to watch a dude get catapulted over the Yazoo River, but if the dude doesn't come back down, that is just alienating. Those of us who can only gravitate resent those who can levitate as well. Thus our beef with birds. Of course, we have no beef with kiwis; kiwis are neither alienating nor inspiring, just dumpy and trudgy and brown. A kiwi being catapulted—now *that* would be a trajectory we'd be comfortable with—not too trudgy, not too inimitable.

After we established the genre, submitted the world to its imperatives, and eliminated all levitation, we would set out some principles. As it is, the world seems quite unprincipled, with its silliness and sepulchers, winsome bunnies, run-over bunnies, drillmasters and goof-offs. With such a free-for-all—with riverine, wolverine, tangerine, and everybody always begging to differ—we cannot for the life of us make out the worldview of the world-maker. Edification eludes us; Anonymous does not appear to have moral designs on anyone. This gives us the willies, in two ways: first, because we like things to have strong messages, strong ideals, even when they do *not* confirm our convictions (then we can write them off), and second, because we don't really like how we come off when we stand next to crocodiles, or loofah plants. We don't like to think of ourselves as prim. We don't like to think of ourselves as propagandists; but next to a tree leaf, what leaf from what book isn't going to look like propaganda?

We should concede that we are not wholly objective, as reviewers; we are not just audience but performers as well; and whenever performers are involved, the execution of a composition is not entirely up to the composer. One may write a stirring oratorio for twenty-three sheep and a donkey and book the finest venues in Oslo and Madrid, but the quality of the performances will depend largely on the choristers themselves, how well they manage their quibbles and tensions, their stage fright, their itchy flanks and twitchy ears; how nimbly-ardently-ruefully they enter the music night after night. Sometimes in the show called "Earth" we see players not playing their parts very nimbly. But there was no rehearsal time, and some of us feel we were miscast, and sometimes our parts seem unperformable: the onus is always on the composer to write a performable piece with a satisfying conclusion.

Which brings us to the last problem we would like to address—the absence of Anonymous. We are assuming he is not around anymore, because we have never seen him posing for his picture in the sky. When a composer dies or absconds before finishing a composition, someone else may step in to complete the rougher sections and notate the last, fragmentary movement. This is called *realization*; Franz Xaver Süssmayr realized the Requiem after Mozart lost his entity. This is what we would do for Anonymous, who seems to have wandered off, leaving everything to ramble on and on—vines making wine, ducks making ducks, the moon sunning itself, the sun mooning around. We've been accused, as a species, of having a mania for conclusion, but it is so imaginatively taxing to sit through a performance that never ends. This one is needlessly endless: for too long we have been denied, by the restive, rambling nature of the world, the keen pleasure of judgment. At every turn, things changed, and our judgment was snatched away, tossed in the water—splash—hard to distinguish from the splashes of little boys having fun. And in exchange we were handed—confusion, sadness. The word "planet" comes from the Greek word *planan*, "to wander"; it is so hard to subject a wanderer to judgment.

We would realize the world; we would wrap it up. It is an artistic pressure; an unfinished work feels like a monkey on our backs. More and more the whole rigmarole seems irrelevant, with its drunken-donkey aimlessness, and willfully wasteful. Anonymous went too far, as young artists tend to do: the sky was overstarred, the ponds overswanned, the evergreens overgreen, and the soul always so overdramatic. Everywhere fools, bums, and blobs were sustained by rank vegetation, rank oxygen, rank sunlight. It was a shame, all that squandered sunshine, all that squandered rain, and the incidental, needless rainbows. We hope that in his sophomore effort

Anonymous will expunge the excesses, tone down that splashy, windy, salty, flighty, flowery, corny, nutty voice, and figure out what he is trying to say. Those damned, dangling, refractory rainbows, availing no one, pertaining to nothing: as if dazzle were enough.

THE BENEVOLENCE OF BLUEBERRIES

"Blueberries support heart health."
"Blueberries support our brain function."
"Blueberries support healthy blood pressure."
"Blueberries support enhanced fat burning."

—The Internet

Blueberries are so supportive. Whoever you are—Vonda the Hothead, Hank the Blank, King Hardecanute—blueberries will support your heart health and your fat-burning and your brain-functioning, and with their anthocyanins they support your longevity: blueberries can help you get very old. This goes for tedious brains and twinklebrains, the brain of the harpist, whose listeners are blissing out, and the brain of the person harping on one subject, whose listeners wish they were tumbling down a mountain. Blueberries also support tumblers down mountains, as well as balanced persons, plowers through life.

Blueberries support you in all your ventures—your picketing, paddling, peddling, pickling, your feeding of ducks, your rinsing of spoons. They support your transitioning into another gender or genre or into another grunge band or your transitioning out of grunge bands altogether—your flaunting of forgettability. You don't have to be famous to be forgettable. Neither do you have to be famous to be supportable: blueberries support the people in the book of *Who's Who* as well as the people in the book of *Who's Through* as well as the people in the book of *Who's Two*—Lila, Otto, Graham, Quinley, and Juniper.

Blueberries are pan-benevolent. They boy the boy and bear the bear and bird the bird. They support the her of your heart, the him of your heart, the hundreds of honeys of hundreds of hearts; they promote the green-eyed girl, the brown-eyed cow, and the whole hedgehog hodgepodge—presumably hedgehogs are as heterogeneous as everyone else. Blueberries are heterogeneously convertible: there goes a businesswoman made of blueberries, a gospeler, a grumbler, a gabbler, and a gambler, all made of blueberries. To the blueberry, no brain is spurious, however meek and mushy its children, or however childless.

Children start out as babies, and babies are hard to understand, but they do come with intimations, like when you hold that one baby and can't help but sing the Raga Bhoopeshwari. Brainchildren start out as babies, too, and some brainbabies seem ridiculous. But since everything is either objectively or subjectively ridiculous, except the stars, this should not be an issue. The stars cannot be your brainchildren. The fate of many brainchildren is sad, being forgotten or suppressed, or hired out as hacks to other brains, when they might otherwise have become two thousand lines of octosyllabic couplets.

Some brains are happy to captain other brains, and some brains are happily captained. Sometimes one pharaonic brain decrees that all offspring of other brains be annihilated, and many brains are not so dismayed to receive ruly little replacements for their ruffians. But maybe one brain weaves a buoyant vessel for her baby and entrusts it to the river, perhaps to float into the hands of a benevolent princess, who supports the baby until it gets big and leads the slaves out of captivity. One princess being benevolent to one baby may result in a mass emancipation.

Between benevolence and malevolence is harmlessness, like pedantry and Ps and Qs, harmless foods and books and customs, all of which proliferate easily, like chickens, and are easily scattered, like chickens. Harmlessness is a land numinous not with deities but with chickens.

Harmlessness is a big stein of nothing. Raise that heavy mug to your lips, drink deep, go on, keep guzzling. The main problem with harmlessness is that it is habit-forming. Since you can never get to the bottom of nothing, you might start knocking it back all day long, and far into the night, far into your longevity.

Yew trees are not harmless, though they resemble Christmas trees, brushy and triangular. But yew trees have no holiday and, unlike Christmas trees, are sometimes surrounded by the limp bodies of cows and pigs and horses. There is death in those berries. Other symptoms of yew poisoning include trembling, drooling, and disappearance of the P wave. The rhythm of the heart has all those lettered waves. Budgerigars may get depressed and persons may lose their color vision. Yew takes the blue from the bird and the you from the pig, sends it all to the nevermind. It also takes the you from the you—yew berries are pan-malevolent. Though you be as illustrious as the moon, you are as poisonable as a pig.

Perhaps you are as illustrious as a pig. Pigs are not illustrious, but they are improbable. Pigs in particular wear their improbability lightly. When the piglet first emerges, everybody thinks, *How improbable!* But then the thought subsides. Once in a while, when piglets are elfing around, we remember how improbable they are, or when they stand next to something probable, like dead piglets or smithereens or apocalypse.

What poison berries do is accelerate the probable. Losing your P wave can affect your Q and R waves and S, T, and U waves. The Ps and Qs of the heart are not insipid but electric. Once the heart is no longer going P-Q-R-S-T-U, P-Q-R-S-T-U, P-Q-R-S-T-U, the brain can't go I W-A-N-T T-A-G-L-I-A-T-E-L-L-E, the brain can't be with child, the brain and heart and corpuscles and ganglia and trapezoids all go back into the sossle. The sossle, made of hotheads, harpists, headbangers, Hula-Hoopers, hajjis, huskies, hucksters, hamsters, and hedgehogs, is nevertheless homogeneous.

But blueberries are agents of heterogeneity, agents of idiosyncrasy, agents of the improbable, agents of the electric. For blueberries to sustain your electric heart with all its electric letters and keep you subject to tagliatelle, keep you discrete, keep you out of the sossle, is such a forgettable phenomenon, like how the trees are always leaving. The trees are forgettably leafy, life is forgettably electric, and there is nothing so forgettably improbable as a piggy, chickeny, humany you. But this is life in the pre-apocalypse—sunny day, moony night, numinous with everyone.

BEASTS IN THE MARGINS

The illustrations in medieval manuscripts started out suitable and pertinent, shinily befitting the text they accompanied. On the first page of a medieval manuscript of Ezekiel, Saint Ezekiel holds a gilded "E." Beside an account of David cutting off Goliath's head, there is David chopping off Goliath's helmeted head. Alongside an account of the shipwreck of Ursula and the eleven thousand virgins, Ursula and the virgins are in a ship going diagonally down into the water. In one manuscript, a man shows up with his liver hanging out, then his intestines, then his heart. He shrugs; he does not seem to be able to keep his organs in. Some body parts can be public or private, as you wish, but other parts are fundamentally private. Even now, in our very public age, the heart is still a private organ.

Since this fellow figures in a surgical treatise on how to put organs back in, he is apposite to the text. All the early illuminations are apposite, as if there were a sign outside the scriptorium: "Only authorized personnel welcome. No weasels, no mermaids, no squabbling human-duck hybrids. If you do not figure in the text, please leave the premises."

But then, in the fourteenth century, the illustrations went off the deep end. In the Luttrell Psalter, in Psalm 7, the text reads, "Let the assembly of the peoples encompass you." *Snail.* "The Lord judges the peoples." *Hedgehog, Weasel.* On other pages, dragonflies hover between the lines. An ape carries her children up the margin, and a unicorn pokes the roses. A man sticks his tongue out at a dragonfly, and someone plays an instrument that is half bagpipe, half human. A monkey chides an owl; another monkey gazes off the page. Here's a blue fool being carried astride a stick, and here's a snake-hipped woman beating a man with her spindle. Here is a man with a blue bouffant and orange monster-heads for feet. A beast with shaggy red paws roars at a lady, who looks at him asquint, as if to say, *What, may I not stand here?* In floats a mermaid.

Who let the weasels into the Psalms, who let the monsters into the psalter? Who forgot to secure the premises? It's as if, after the monks had gone to bed one night, a bunch of pranksters crawled through the windows of the scriptorium and tramped all a-silly onto the sacred pages. Even taking into account that these were silly times, when a duck might say "queck" instead of "quack," and instead of being thanked by a lady, you might be thonked, these illuminations seem awfully off-topic.

The question is, can the Psalms handle such eccentric marginalia? Most songs would be upstaged if you wrote down their lyrics and a snail crossed the page; swaddled babies would steal the show, as would twerps bonking each other on the head or monsters scarfing down their own necks. What poem isn't going to be discombobulated by a monkey who sits down on the page and gazes *off* the page? It takes a special song to share the page with uninterested monkeys, or rowboats, or snails with pizzazz: it takes a song with which nothing can be incongruous. Songs of indoctrination, songs of advertisement wouldn't swing it: for the sake of clarity, and credibility, and convincement, they have to ditch the weirdos. Snails'll get the ax.

Actually, there are four kinds of songs that might

countenance snails: first, songs about snails; second, songs of jibber-jabber—with which nothing can be incongruous *or* congruous; third, songs that comprehend every blinking thing in the universe, including gastropods, gluons, the Blinking Planetary, "the secrets of the heart." The last, and most feasible, kind of song that will countenance a snail is a song of questions—because a question will countenance anything, as a window will. *Who can keep alive his own soul? Who can understand his errors? What is man that you are mindful of him? How long, O Lord?*

Old questions are so intoxicating, like old grapes. Of course, for the sake of accomplishment, we can't be gazing out the window all day, contemplating questions. Proactivity closes the curtains and sweeps the house and organizes the winter gear. Proactivity is pretty good at getting the mittens to shape up, though it is not sure what to do with ungovernable things, like mermaids and yetis and things to come and who you like and who you don't like and what you dream about. Dreams are perverse and freewheeling and they're always disregarding your tenets. You can think about politics all day but then dream you are walking all alone in a forest of cedar trees. You can believe you've finally firmed up your identity but then dream you are three people riding three bicycles. Once, I dreamed that someone reached into his chest, pulled out his heart, and handed it to me.

But then is life really so much more governable than dreams? "*But alday falleth thyng that foles ne wenden,*" Chaucer wrote. All day things happen that folks did not expect. All the spoons in the house disappear, or you've just started thinking an interesting thought when somebody throws a clod at your head. Once, the toddler told me there was a kangaroo in the yard, and I looked and lo, there was no kangaroo. Once, the toddler said there was a parrot in the yard, and I looked out the window, and there was a parrot, big and red and blue, perching on the clothesline in our backyard in Montana. Some animals are fictitious and some are not. I know someone who was driving up a mountain road and a bear fell out of a tree onto her Mini Cooper.

Once, it had rained all day, so the doors in our house swelled up, so when I was giving the children a bath the door got stuck, and when I tried to yank it open, it just got stucker and stucker, trapping the three of us in the bathroom. I yelled for help out the window, but since it was still raining, all the neighbors were inside, except for the little brown birds. Little brown birds are such flittery neighbors.

Lord, who may abide in Your tabernacle? Who may dwell in Your holy hill? Heaven only knows about the holy hill, but who may dwell on the earthly hill is plain to see: little brown birds, unwonted parrots, jokers, smokers, the highborn, the hayborn, "every beast," people who look like calabashes, people who feel like worms, people who are not into universalism. The world appears to be illustrated with as much drollery as the Luttrell Psalter: maybe those wacko illuminations were actually a naturalistic turn. After all, sometimes when one is muttering the Psalms to herself, she may cross paths with a snail. Or, just as one was thinking about *me and my salvation*, there's a weasel, yo.

Who may dwell on Earth seems similar to who may dwell in the fourteenth-century psalter: lots and lots of unauthorized personnel. In the Middle Ages, if you had a loathsome disease you were supposed to clack your castanets so no one would come near you, and you had to stay far away from the tournament tents. But *in the heavens God has pitched a tent for the sun.* A tent big enough for the sun and moon to stroll around in is big enough for the germiest of persons. The germiest, the squirmiest, the squabbliest, the most incongruous. *He that sitteth in the heavens shall laugh.* Looking at us there, in the margins of the Psalms, squinting at monsters, japing at dragonflies, climbing plum trees in our bare feet, trying so hard to tune our harps, committing spindle crimes, turning halfway into ducks and bagpipes, we see how laughable we were, how saucy-tiny-bonky-beamy, how fully unadvised of things to come, how incapable of keeping alive our own souls, how someone might love us with a fever.

GLOSSARY

ADORAZZI: The big difference between the *adorazzi* and the *paparazzi* is that the *paparazzi* are discriminating in their adoration, whereas the *adorazzi* are promiscuous. They have no standards, no principles; they are sweet on even thistles and pigs.

AIR PLANTS: Air plants are used to make air guitars, and you can hear that in their song. But violins, when they were maples, dug down deep, and of course tubas were born deep.

"ALICE IN WONDERLAND": A song not madcap but melancholy, like how I feel when I think about little girls fallen into a land full of bossy authorities wielding arbitrary rules, little girls who will never climb back out.

BLOBFISH: When a blobfish shined its blobby countenance on us, we felt honored and favored and also giggly.

BLUE FOOL: Blue is an unusual color for fools. Fools are usually pink like pigs or brown like moose. Moose are almost never blue. Once in a blue moon is rare but not as rare as once in a blue moose.

THE BOOK OF JOB: The book of Job is a book of yore. Books of yore are placed all together on one little shelf in the back of the bookstore. However, there was not just one yore but many yores, and in that aspect yores are like yonders.

BRIEFLY A BABY: All babies are tight for time, even if their century is not yanked out from underneath them. First they learn to chatter and totter and honk the horn, and then they go through an unhinged period, after which they get more and more hinged until they want to be investors.

BROTHERS AND SISTERS: Of course with siblings there will inevitably be frictions. Brother Juniper is always needling us and we are always ruffling Sister Chicken's feathers.

CATS AND CHURCHES AND TREES: Many obnoxious cats come and go, many obnoxious churches stick around for a while, and trees tend not to be obnoxious.

CHACHALACAS: Like screamo singers, chachalacas are hard to lose track of. Have you ever heard anyone going around saying, *I can't find the chachalacas*? I didn't think so.

ELECTRICAL SPRITES: At any neighborhood park, you can watch electrical sprites flying in the face of convention. Don't be deterred if they are playing baseball or football—if preschoolers are playing them, sports are actually fun to watch. Baby sports are good too. People tired of raising their babies should try racing their babies sometimes.

ELVERS: Baby eels. Elvers have nothing to do with elves.

FRANÇOIS LEGUAT: In the 1920s, one Mr. G. Atkinson accused Leguat of having invented the solitaire, and also of having never existed. Most fiction writers just invent their material, but Leguat invented his material *and* himself *and* he was a good guesser, since his fictional solitaires, with their little wings and big wing knobs, exactly match the nonfictional skeletons found in caves on Rodrigues Island.

"GO TO THE DEVIL AND SHAKE YOURSELF": A tune also known as "When Sick Is It Tea You Want?" which seems more solicitous.

GOG AND MAGOG: Gog and Magog are odious men or wicked

lands or progenitors of the Scythians or apocalyptic hordes or, as Gogmagog, a single British giant. Whatever they are, they seem to be something to avoid.

GOOD DOGS AND BAD DOGS: I wonder if there are good squids and bad squids, too. It seems that in any profession—poetry, podiatry, cheesemongering—there are both good and bad practitioners, and that usually the good ones are more successful. However, when it comes to prophecy, the reverse is true. Bad prophets tell people what they want to hear, so they have nice new offices, whereas good prophets get thrown into cisterns. If you are looking for a good prophet, go peer into some cisterns and you might see one sinking down into the mud.

A GRAPE UPON WHOM NOTHING IS LOST: Once, in a women's magazine, I read about somebody upon whom everything was lost. Even though she was in her sixties, this woman had enviably smooth, unlined skin, all because for the past thirty-five years she had been in a coma.

GREAT DISMAL SWAMP: The Great Dismal Swamp attracts otters and cypresses and loblollies and inkberries and fishermen and tupelos and stinkpots and bobcats and boaters and bullfrogs, but not many brides.

"I LOVED BEING OLD": In my house lives a toy rabbit holding a basketball, who—I am told—likes everything except for death. He is friends with a macaroni penguin who likes everything including death. She is the kind of penguin who would not only say, *I loved being young* and *I loved being old*, but also *I loved being dead.*

INKA TAKY TRIO: Consisting of a cousin and a husband and the Incan princess Zoila Augusta Emperatriz Chávarri Del Castillo. When she was young, Princess Zoila Etc. cultivated the shocking range of her voice by singing to rocks on Peruvian mountains. If you wish to cultivate a shocking voice, rocks are a good first audience, because rocks are hard to shock.

INTOXICATING OLD QUESTIONS: If old questions are like old grapes, old answers are like old nuts. Go ahead and stay sober on your rancid nuts if you want.

KING HARDECANUTE: King Hardecanute and King Harold were co-kings but maybe not copacetic: after Harold died, Hardecanute had his remains disinterred and flung into a fen.

MASERATI: If flowers could get hold of a Maserati, they would turn it into a big shiny flowerpot.

MELANCHOLY WOODPECKERS: This is their name, not their

temperament, though I'm sure they *have* temperaments. The sanguine-phlegmatic melancholy woodpecker should think twice before marrying a melancholy-choleric melancholy woodpecker.

MISTAKES MADE IN SNOW: This is in contrast to mistakes made in plastic. Snow is a more forgiving medium.

MOBOCRACIES: The mobocracy may be crazy but you can be crazy without a mobocracy, too.

MYSTERY: For some people, mystery's an anachronism, an extraneous holdover from times past, like doilies.

ODD DUCKS: You may have known a few odd ducks in high school, among all the normal ducks. The term becomes problematic when you try to apply it to ducks, though, because all ducks are odd.

OTHER EXTRA-NATIONAL GIFTS BESIDES THE TIGER: Boxes of sunshine, bags of grass, bolts of lightning tied together with twine, a bucket of rain and a basket of stars, and a bunch of jumpy little birds.

OTHER VIRTUES BESIDES RELEVANCE: For example, phosphorescence, fizziness, funniness, and bunniness.

THE PATIENCE OF THE SANDHILL CRANE: She is patient, she is stoical, she does not laugh at our jokes. Of course, it is not necessarily true that someone is stoical if she does not laugh at our jokes. It could be that our jokes are not funny enough.

PENGUINS TOO RETICENT: The penguin is not reticent; the penguin is asleep.

"THE PERSECUTION OF THE GOLDFISH": Although in her bowl on top of the piano Miss Goldfish appears demure, goldfish in the sea are cruel and autocratic. You should see how they work the other fishes over.

PIANO CRIMES IN ONE'S VERY OWN APARTMENT BUILDING: As the woman who lived in the apartment downstairs from Igor Stravinsky reported, "Monsieur Stravinsky plays only wrong notes."

POMERANIANS EXCLUDED FROM YELLOWSTONE: Although they are excluded from many national parks, Pomeranians do have a club, and it's not every animal that gets a club. Pomeranians have a club but they don't get to belong to it.

PRECURSORS: Extinct animals were failed precursors.

PROPOSED EXISTENCE: The Oort cloud is an example of something with a proposed existence. I can't think of an example of something whose existence has not yet been proposed.

PROVINCIALISTS: Provincialists say to their children, "I didn't raise you to be an astronomer."

RAGA BHOOPESHWARI: When you listen to ragas they color your brain bumblebee yellow or boysenberry purple or sangria red or parakeet green. When you do not listen to ragas your brain is brain gray.

SIMOOM: Blistering sandy wind that frizzles you and your camels. Once, I lived in a basement apartment that was heated by a simoom minus the sand. The apartment was icy cold except for when the furnace roared hot air through the big square hole in the ceiling and desiccated me and my camels.

SOGGY WORDS: Soggy words are words without thoughts or words in federalese or words spoken to inspire folks to buy things. They are particular to human beings. Swamp rats do not know the anguish of having spoken soggy words.

SONGS ABOUT SNAILS: Since there aren't many songs about snails out there, if you want to sing about snails you can either come up with your own original lyrics or you can just take songs about "saints" and substitute the word "snails." *I sing a song of the snails of God . . .*

SOSSLE: An obsolete word for obsolete material, like dead donkeys and dead dairymaids. Actually, to be obsolete means that you were once in fashion, so some donkeys and dairymaids will never be obsolete.

SWALLOWED BY A CATEGORY: Of all the hungry predators chasing humans around, the hungriest may be the categories, whose hunger is never diminished but increases with every bite. Some people run away from them screaming but others jump with joy right into their ravening mouths. But categories never consult any of their members. Once you've been swallowed by the Formalist or Thermidorian category, your thoughts are not your own but Formalist thoughts, Thermidorian theories.

TAXON: If you are reading this, then your taxon is *Homo sapiens*. However, one is more than one's taxon. If the science on humans is incomplete, the science on Sid Vicious is even more incomplete.

TRIANGULAR GUY IN THE SKY: You will know him by his triangularity. A triangle is smaller than a circle but larger than a square.

TYCHO BRAHE'S MOOSE: Before his moose fell down the stairs, Tycho Brahe would lend him out to nobles to have at their parties. If someone started getting tedious, the noble could say, *Have you met Merle, my sometimes moose?*

UNIVERSALISM: Universalism is a religious ideology that holds that all humans will be saved, whatever their sect or non-sect. My problem with universalism is that it is too exclusive, so I have made up an ideology that includes not just all the human rascals but also all the buffalo rascals and reptile rascals and paddlefish and turkeys and centipedes and wombats and warty pigs and I call it everybodyism.

WANDAS AND WAYNES: I have known three Waynes in my life and they were all quality Waynes. I have known one Wanda and she was a quality Wanda. I have never known a Wilma or a Flossy.

***WHY WON'T THEY LISTEN?*:** An actual pamphlet, an actual attitude.

YAN, TAN, TETHER, METHER, PIP: Yan, tan, tether, mether, pip, azer, sezar, akker, conter, dick, yanadick, tanadick, tetheradick, metheradick, bumfit, yanabum, tanabum, tetherabum, metherabum, jigget. Last summer there were bumfit poppies growing in my yard and this year I was expecting at least yanabum or jigget poppies, but then only pip poppies bloomed, since someone snapped the big green buds off their stems to give as treasures to her big brother.

ZEUS: Zeus was an accomplished god who sent dreams to Greek people and swallowed his wives and threw mountains at monsters but not anymore because Zeus vamoosed. (Zeus vamoosed from this book too because I removed the essay in which he appeared.)

ZIGZAG BRIDGES: Since devils can travel only in straight lines, once you've zigged and zagged your way to the other side of the stream, you will have ditched all those stricty-pants little devils who were following you. Olé!

NOTES

IN LIEU OF A WALRUS

42 "The mule I sit on . . .": *The Gift: Poems by Hafiz the Great Sufi Master*, trans. Daniel Ladinsky (New York: Penguin Compass, 1999), 133.

NON SEQUITURS

49–53 Passages quoted in this chapter are from the King James Version and from *The Jewish Study Bible* (New York: Oxford University Press, 2004).

GLOSSARY

185 "Monsieur Stravinsky . . .": Quoted in Jan Swafford, *The Vintage Guide to Classical Music* (New York: Vintage, 1992), 409.

ACKNOWLEDGMENTS

Thanks to the Ragdale Foundation for the five days to write and all that wonderful food, and thanks to Mary Margaret Alvarado for the further five days to write and all that wonderful food. Thanks to Eula Biss, Sean Hopkinson, Steve Marty, Kerry Reilly, Sharon Leach, and Lauren Bongard Schwartz for reading this book when it was still green.

Thanks to Glover Wagner for the words of life, to Alanna Brown for the words of encouragement, to Ingrid Stuart and Jody Hechtman and Carrie Fry and Eric Beavon for unwavering friendship.

Thanks to Jin Auh for elegantly shepherding my work, to Jenna Johnson for befriending my book and shedding so much light on it, and to Yiyun Li for egging me on for eighteen years.

Thanks to my parents, Benjie and Sharon Leach, for hanging the moon and helping me in millions of ways. Thanks to my in-laws, Lorna and Rick Lukens, for frequently driving to Montana, as not all patrons of the arts are willing to do, to provide child care.

Thanks to Peter and Sylvie for being children of enormous, electrifying influence, and thanks to Matthew for bearing all things, believing all things, hoping all things.

A NOTE ABOUT THE AUTHOR

Amy Leach grew up in Texas and earned her MFA from the Nonfiction Writing Program at the University of Iowa. Her work has appeared in *The Best American Essays*, *The Best American Science and Nature Writing*, and numerous other publications, including *A Public Space*, *Orion*, *Tin House*, and the *Los Angeles Review of Books*. She is a recipient of a Whiting Writers' Award, a Rona Jaffe Foundation Award, and a Pushcart Prize. Her first book was *Things That Are*. Leach lives in Montana.